Question Bank on ENTOMOLOGY

for ARS, NET, JRF, SRF, UPSC, SAUs and Allied Exams.

The Author

Dr. Supriya Biswas is devoted in Entomology for previous 10 years. He passed B.Sc. (Ag.) from Post Graduate College, Ghazipur (affiliated to the VBS Purbanchal University, Jaunpur) and M.Sc. (Ag.) in Agricultural Entomology with the University Merit Scholarship, from Uttar Banga Krishi Viswavidyalaya, Pundibari, Cooch Behar, West Bengal. He completed Ph.D in Agril. Entomology (Major in Economic Entomology) from Uttar Banga Krishi Viswavidyalaya. He has passed NET, 2014 Examination conducted by Agricultural Scientists Recruitment Board (Indian Council of Agricultural Research) New Delhi. Dr. Biswas has published 27 Research Papers, 02 Book chapters and 04 popular articles in different national and international journals. He has the distinction of having attended and presented papers at several national and international seminars, symposium and workshop. Dr. Biswas also has membership of different national and international scientific societies. The author has 6 years of experience on Teaching (Both UG and PG) and Research at a century old reputed college and he has served four Universities. Presently Dr. Biswas serving the Department of Agricultural Entomology, Uttar Banga Krishi Viswavidyalaya, Pundibari, Cooch Behar, West Bengal.

Question Bank on ENTOMOLOGY

for ARS, NET, JRF, SRF, UPSC, SAUs and Allied Exams.

Supriya Biswas

2016
Daya Publishing House®
A Division of
Astral International Pvt. Ltd.
New Delhi – 110 002

Cataloging in Publication Data--DK
Courtesy: D.K. Agencies (P) Ltd. <docinfo@dkagencies.com>

Biswas, Supriya, author.
Question bank on entomology : for ARS, NET, JRF, SRF, UPSC, SAUs and allied exams. / Supriya Biswas.
pages cm

ISBN 978-93-5130-968-0 (International Edition)

1. Entomology--Examinations, questions, etc. I. Title.

QL463.B57 2016 DDC 595.7 23

Published by : **Daya Publishing House®**
A Division of
Astral International Pvt. Ltd.
– ISO 9001:2008 Certified Company –
4760-61/23, Ansari Road, Darya Ganj
New Delhi-110 002
Ph. 011-43549197, 23278134
E-mail: info@astralint.com
Website: www.astralint.com

Laser Typesetting : **Classic Computer Services**, Delhi - 110 035

Printed at : **Sanat Printers**

— Dedicated to —

my beloved son
Vivaan
&
my teachers

Who made me capable for this work

Preface

In present days competitive examinations have become a routine affair. Most of the Universities, organizations and institutes have adopted entrance examination. To conduct such type of examinations mostly objective type of questions are followed. This requires quick answering within a specified period of time to secure high merit and assured success. While appearing for examinations and stimulating students for examination I felt lack of study materials and that there should be a book on the complete syllabus of Entomology.

This book is framed holding in view the pattern of various competitive examinations. In this book wide area has been covered chapter wise with student friendly easy language, recent data about some important features have been updated and egg laying, pupation behaviour of important crop pests have been compiled in tabular form. I feel after following the book thoroughly, students shall stride into the examinations with clear concept, confidence and ultimately will become the winner. Professionals of Entomology may use this book as a ready reference.

I affectionately articulate my thankfulness to my parents and all my family members, their constant love and affection gave me the self-confidence and dare to dream. I wish to express my thanks also to my wife, Jui for her unstaiting help in all ways.

It is truly an immense chance for me to pay my acknowledgment to all my teachers from infancy, who have imbibed in me the thirst for knowledge. I express my gratefulness to Subhajit (Islampur), one of my students, who started the painstaking typing of this manuscript. I also extend my thankfulness to Dr. Arup Sarkar, Prof. Asish Kumar Singha Roy, Dr. Biplab Mitra, Dr. Deb Sankar Gupta, Dr. Debapriya Sarkar, Prof. Debasis Mazumdar, Debojyoti Dutta, Dr. Indrajit Biswas, Dr. Joydeb Ghosh, Ms. Moultia Chatterjee, Dr. Nilanjana Choudhury, Dr. Nilay Ray, Dr. Nripendra Laskar, Dr. Partha Sarathi Patra, Mr. Prasad Singh, Prof. Sabita Kumar Senapati, Dr. Samsul Haque, Dr. Subhendu Kumar Roy, Dr. Sunil Kumar Ghosh, Dr. Suprakash Paul, Dr. Surajit Khalko, Dr. Swapan Kumar Barman and Prof. Tapan Kumar Hath.

My special thanks are also extended to Dr. Ajit Kumar Roy, Medical Officer, Uttar Banga Krishi Viswavidyalaya for giving the correct suggestions and helped me during writing the chapter Pests of Medical Importance.

I am heartily grateful to the publisher of this book whose sincere effort made it available to you in a short time. I hope the book would meet the requirements of the students and professionals of Entomology, and their valuable suggestions would be most appreciated.

Supriya Biswas

Contents

Syllabus of Agricultural Entomology for NET and ARS Examination of ASRB (ICAR)

Unit 1: Systematics

History and development of Entomology, Evolution of insects, position of insects in the animal world, characteristics of phylum Arthropoda, structural features of important arthropod groups such as Trilobita, Chelicerata and Mandibulata, structural features of important classes of phylum Arthropoda *viz.* Arachnida, Crustacea, Chilopoda, Diplopoda and Hexapoda. Classification of insects up to order level, habits, habitats and distinguishing features of different Order and important Families.

Unit 2: Morphology

Body wall, its structure, outgrowths, endoskeleton, Body regions, segmentation, sclerites and sutures. Head and head appendages, types of mouth parts, antennae, their structure and types. Thorax structure, thoracic appendages and their modification. Wings, their modification and venation, Abdomen; structure, abdominal appendages both in Pterygota and Apterygota. External genitalia, general structure and modification in important insect orders.

Unit 3: Embryology, Internal Anatomy and Physiology

Embryonic and post embryonic development, types of metamorphosis, physiology of ecdysis. General features and types of larvae and pupae. Structure, function and physiology of Digestive, Circulatory, Respiratory, Reproductive, Nervous and Excretory systems, Sense Organs; structure and types. Insect food and nutrition; minerals, carbohydrates, proteins and amino acids, lipids, vitamins and their role in growth and development, artificial diets.

Unit 4: Ecology

Concept of ecology, Environment and its components-biotic and abiotic factors and their effects on growth, development, population dynamics, distribution and dispersal. Principle of biogeography and insects biodiversity. Biotic potential and environmental resistance. Ecosystems, agroecosystems analysis, their characteristics and functioning. Intra and inter specific relationship; competition, predator-prey and host-parasite interactions, ecological niche. Life table studies, population models. Food chain and food web. Arthropod population monitoring, pest forecasting. Diapause and causes of pest out breaks.

Unit 5: Biological Control

Importance and scope of biological control, history of biological control: Biocontrol agents-parasites, predators and insect pathogens. Important entomophagous insect Orders and Families. Ecological, biological, taxonomic, legal and economic aspects of biological control, phenomena of multiple parasitism, hyperparasitism, superparasitism and their applied importance. Principles and procedures of using exotic biocontrol agents. Utilization of natural biocontrol agents: conservation, habitat management and augmentation. Mass multiplication techniques and economics. Effective evaluation techniques, Biocontrol organizations in world and India. Successful cases of biological control of pests. Entomophilic pathogens: bacterial, fungi, viruses, rickettsiae, Protozoan and nematodes, Modes of transmission, methods of uses, symptoms of infection. Microbial insecticides and their formulation. Merits and demerits of microbial control. Role of biocontrol agents and microbial insecticides in Integrated Pest Management.

Unit 6: Chemical Control and Toxicology

History, scope and principles of chemical control. Insecticides and their classification. Formulations of insecticides. Susceptibility of insects to the entry of insecticides. Physical, chemical and toxicological properties of different groups of insecticides: chlorinated hydrocarbons, organophosphates, carbamates, synthetic pyrethroids, chlordimeform, chitin synthesis inhibitors, avermectins, nitroguandines, phenylpyrrozzoles, botanicals (natural pyrethroids, rotenone, neem products, nicotine, pongamia spp. etc). Combination insecticides. Problems of pesticide hazards and environmental pollution. Safe use of pesticides, precautions and first aid treatments. Insecticides Act 1968, registration and quality control of insecticides. Evaluation of toxicity, methods of toxicity testing, determination of LD 50, LT 50, RL 50 etc. Pesticides residues in the environment and their dynamics of movements, methods of residue. Pharmacology of insect poisons. Mode of action of different groups of insecticides; neuroactive (axonal and synaptic) poisons, respiratory poisons, chitin synthesis inhibitors. Metabolism of insecticides; activative and degradative metabolism, detoxification enzymes and their role in metabolism. Selectivity of insecticidal actions; insecticide resistance; mechanism, genetics and management of insecticide resistance.

Unit 7: Host Plant Resistance

Chemical ecology: mechano and chemo receptors. Host plant selection by phytophagous insects. Secondary plant substances and their defenses against phytophagous insect. Basis of resistance (Antixenosis, Antobiosis, Tolerance). Biotypes development and its remedial measures. Tritrophic interactions, induced resistance. Breeding for insect resistant plant varieties. Resistance development and evaluation techniques. Genetics of Resistance: vertical resistance, horizontal resistance, oligogenic resistance, polygenic resistance. Biotechnological approaches and development of transgenic insect resistant plants, its advantages and limitations. Case histories. Insect resistance to transgenic plants and its management.

Unit 8: Innovative Approaches in Pest Control

Behavioral control: pheromones-types and uses, advantages and limitations. Hormonal control: types and function of insect hormones, insect hormone mimics, advantages and limitations. chemosterilants, antifeedants, attractants, repellents; their types, method of applications, advantages and limitations. Genetic control: concepts and methods, case histories, advantages and limitations. Potentialities of IPM.

Unit 9: Integrated Pest Management

History, concept and principles of IPM. Components of IPM: Host plant resistance, agronomic manipulations, mechanical and physical methods, chemical methods, biocontrol agents utilization, genetic and behavioral control strategy etc. IPM strategies for field and horticultural crops. IPM case histories. Concept of damage levels- Economic threshold levels (ETL), Economic injury levels (EIL) and their determination. System approach, Agro ecosystem and cropping system *vs.* IPM. Constraints and Strategies of IPM implementation.

Unit 10: Pesticide Application Equipments

Types of appliances: sprayers, dusters, fog generators, smoke generators, soil injecting guns, seed treating drums, flame throwers, etc. Power operated sprayers and dusters. Types of nozzles and their uses. Maintenance of appliances. Aerial application of pesticides, principles of aerial application, factors affecting the effectiveness of aerial application. Equipments for aerial applications. Advantages and disadvantages of aerial application.

Unit 11: Pests of Field Crops and their Management

Distribution, host range, biology and bionomics, nature of damage and management of arthropod pests of cereals, Oilseed, pulses and fibre crops, sugarcane and tobacco. Polyphagous pests: locusts, termites, hairy catepillars, cut worms and white grubs.

Unit 12: Pests of Horticultural Crops and their Management

Distribution, host range, biology and bionomics, nature of damage and management of arthropod pests of vegetables, fruits and plantation crops, spices, condiments and ornamentals.

Unit 13: Pests of Stored Products and their Management

Fundamentals of storage of grains and grain products. Storage losses, sources of infestation/infection, factors influencing losses, insect and non-insect pests, their nature of damage and control. Microflora in storage environment and their control. Storage structures, bulk storage and bag storage, their relative efficacy and demerits. Grain drying methods and aeration. Non-insect pests (rodents, birds, mites) of stored products and their control. Integrated management of storage pests.

Unit 14: Arthrpod Vectors of Plant Diseases

Common arthropod vectors *viz.*, aphids, leaf hoppers, plant hoppers, whiteflies, thrips, psylids, beetles, weevils, flies, bees and mites and their relationship with the plant pathogenic fungi, bacteria, viruses, mycoplasma. Mechanism of pathogen transmission : Active mechanical transmission, biological transmission. Toxicogenic insects, mites and phytotoxemia. Some important arthropod vector transmitted diseases and their epidemiology in India. Management of vector and its effect on control of diseases.

Unit 15: Honey Bees and Beekeeping

Honey bees and their economic importance. Bee species, their behaviour, habit and habitats. Beekeeping: bee pasturage, hives and equipments, seasonal management. Bee enemies including diseases and their control.

Unit 16: Silkworms and Sericulture

Silkworm species, their systematic position and salient features. Rearing techniques of mulberry-muga-eri and tassar silkworms. Nutritional requirements of silkworms. Sericulture: rearing house and appliances, silkworm breeds, principles of voltism and nioultism, seed production and its economics. Enemies and diseases of silkworms and their management. Sericulture organization in India.

Unit 17: Lac Insect

Lac insect, its biology, habit and habitats. Host Trees: pruning, inoculation, lac cropping techniques, and harvesting. Enemies of lac insect and their control.

Unit 18: Other Useful Insects

Pollinators, biocontrol agents of weeds, soil fertility improving agents, scavengers. Use of insects and insect products in medicines. Usefulness of insects in scientific investigations, insects as food.

Unit 19: Statistics and Computer Application

Frequency distribution, mean, mode and median. Standard, normal, bionomial and Poisson's distribution, Sampling methods and standard errors. Correlation and regression: Partial and multiple, tests of significance; t, F, chi- square, Duncan's multiple range tests. Design of experiments: Principles of Randomized block design, Completely randomized block design, Latin square design, Split-plot designs. Probit analysis. Use of soft ware packages like SPSS, SAS, etc. for the above tests and designs of experiments for analysis.

Chapter 1
Systematics and Taxonomy

Multiple Choice Questions

1. **The word Entomology belongs to which language**
   (*a*) French (*b*) Greek
   (*c*) Latin (*d*) English

2. **At the present time, scientists have named approximately__________species.**
   (*a*) 100,000 (*b*) 500,000
   (*c*) 1,000,000 (*d*) 1,500,000

3. **Arthropods are**
   (*a*) Entero coelomates (*b*) Pseudo coelomates
   (*c*) Gastro coelomates (*d*) Schizo coelomates

4. **Arthropods evolved during**
   (*a*) Devonian period (*b*) Carboniferous period
   (*c*) Cambrine period (*d*) None of these

5. **According to Imms (1952) total numbers of orders of class Insecta is**
   (*a*) 28 (*b*) 29
   (*c*) 30 (*d*) 31

6. **Number of Apterygote orders**
   (*a*) 9 (*b*) 4
   (*c*) 16 (*d*) 29

7. **Number of Endoptegote orders**
   (a) 9 (b) 4
   (c) 16 (d) 29

8. **Numbers of Exopterygote orders**
   (a) 29 (b) 30
   (c) 16 (d) 9

9. **The order Odonata includes**
   (a) May flies (b) Dragon flies
   (c) Stone flies (d) House flies

10. **Hyper metamorphosis is seen in the insect family**
   (a) Chrysomelidae (b) Melloidae
   (c) Tenebrionidae (d) Cicindelidae

11. **In Hymenoptera, fore tibia with 2 apical spurs is a characteristic feature of the family**
   (a) Tenthredinidae (b) Ichneumonidae
   (c) Chalcidae (d) Braconidae

12. **The largest insect order is**
   (a) Lepidoptera (b) Diptera
   (c) Thysanoptera (d) Coleoptera

13. **The ancient insect order is**
   (a) Neuroptera (b) Strepsiptera
   (c) Orthoptera (d) Collembola

14. **Which of the following order contains most of the parasitic entomogenous insects?**
   (a) Hemiptera (b) Coleoptera
   (c) Strepsiptera (d) Hymenoptera

15. **Which insect orders contain species that are important pests of domestic animals?**
   (a) Siphonaptera and Diptera (b) Orthoptera and Hemiptera
   (c) Coleoptera and Phthiraptera (d) All of these

16. **Suppose you find an intersecting arthropod living on the bottom of a fresh water pond. It has eyes, antennae, mandibles and 10 no. of legs. You conclude that this organism belongs in the class**
   (a) Insecta (b) Diplopoda
   (c) Crustacea (d) Arachnida

17. Which of the following does not belong to the family Gelechiidae?

(a) *Phthorimaea operculella*
(b) *Sitotroga cerealella*
(c) *Holcocera pulverea*
(d) *Liriomyza trifoli*

18. In Lepidoptera, forelegs are reduced and vestigial, folded on the thorax is a characteristic features of the family

(a) Nymphalidae
(b) Noctuidae
(c) Lymantridae
(d) Gelichiidae

19. Which order is exclusively herbivorous?

(a) Hemiptera
(b) Phasmida
(c) Odonata
(d) Thysanoptera

20. Which class is most closely related (Phylogenetically) to the Insecta?

(a) Xiphosura
(b) Crustacea
(c) Myriapoda
(d) Arachnida

21. Which group of insects has simple metamorphosis?

(a) Fleas
(b) Flies
(c) Beetles
(d) Bugs

22. Hind tibia with double row of spines is a characteristic features of family

(a) Delphacidae
(b) Cicadellidae
(c) Membracidae
(d) Cercopidae

23. Pronotum is prolonged backwards into a distinct elevated hood or horn lying over the abdomen is a characteristic features of family

(a) Delphacidae
(b) Cicadellidae
(c) Membracidae
(d) Cercopidae

24. Which order is both ectognathous and wingless?

(a) Collembola
(b) Thysanura
(c) Phthiraptera
(d) Diplura

25. Which insects do not feed on plants?

(a) Thrips
(b) Lacewings
(c) Whiteflies
(d) Sawflies

26. Cornicles are found in

(a) Aphididae
(b) Cicadellidae
(c) Delphacidae
(d) Melanagromyzidae

27. Fly larvae (maggots) move away from a bright source of light. This is an example of

(*a*) Taxis (*b*) Kinesis
(*c*) Reflex (*d*) Transverse orientation

28. Megaloptera and Palnipennia are the suborders of the insect order

(*a*) Neuroptera (*b*) Odonata
(*c*) Trichoptera (*d*) Ephemeroptera

29. Social Insects are belongs to which of the following groups

(*a*) Diptera and Hemiptera (*b*) Protura and Diplura
(*c*) Hymenoptera and Isoptera (*d*) None of these

30. Ability to roll as protective mechanism is seen in

(*a*) Millipedes (*b*) Centipedes
(*c*) Crabs (*d*) All of these

31. Green glands are present in

(*a*) Arachnids (*b*) Crustacians
(*c*) Diplopoda (*d*) Chilopoda

32. Class Insecta (Hexapoda) belongs to the subphylum

(*a*) Trilobita (*b*) Chelicerata
(*c*) Crustacea (*d*) Uniramia

33. Which of the following order showing more Arthropod characters?

(*a*) Protura (*b*) Collembola
(*c*) Diplura (*d*) Thysanura

34. Insect have existed on this earth__________million years.

(*a*) 100 - 200 (*b*) 200 - 225
(*c*) 500 - 600 (*d*) 250 - 500

35. The extinct class of the phylum Arthropoda is

(*a*) Trilobita (*b*) Onychophora
(*c*) Diplopoda (*d*) Chilopoda

36. In Isoptera, a frontal gland (in head) is present in

(*a*) Queen (*b*) King
(*c*) Soldiers (*d*) Workers

37. New addition to the already existing insect orders

(*a*) Phasmida (*b*) Mandophasmatoidea
(*c*) Neuroptera (*d*) Zoraptera

38. Earliest record of Indian insects

(*a*) Systema Naturae
(*b*) Termites of Tanjore District
(*c*) Entomologia systematica
(*d*) Indian Insect Life

39. First Entomologist to the Govt. of India

(*a*) HM Lefroy
(*b*) Fletcher
(*c*) Fabricious
(*d*) Lionel de Niceville

40. First Plant Protection Advisor to the Govt. of India

(*a*) HS Pruthi
(*b*) HM Lefroy
(*c*) Ananthakrishnan
(*d*) S Pradhan

41. Physogastric abdomen is present in

(*a*) Termite queen
(*b*) Grasshopper
(*c*) Cockroaches
(*d*) All of these

42. In Isoptera, the winged forms are

(*a*) Soldiers
(*b*) Workers
(*c*) Both (*a*) & (*b*)
(*d*) Primary reproductives

43. In Isoptera, the sterile forms are

(*a*) Soldiers
(*b*) Workers
(*c*) Both (*a*) & (*b*)
(*d*) Primary reproductives

44. The Entomological Society of India was established in

(*a*) 1938
(*b*) 1948
(*c*) 1958
(*d*) 1968

45. Insects belongs to which of the following order cannot flex their wing over the abdomen?

(*a*) Odonata
(*b*) Ephimeroptera
(*c*) Neuroptera
(*d*) Both (*a*) & (*b*)

46. Caddis fly belongs to the order

(*a*) Collembola
(*b*) Thysanura
(*c*) Trichoptera
(*d*) Psocoptera

47. May flies belongs to the order

(*a*) Odonata
(*b*) Ephimeroptera
(*c*) Plecoptera
(*d*) Thysanoptera

48. Author of Indian Insect Life is

(*a*) Lefroy
(*b*) Hampson
(*c*) Rothney
(*d*) Pruthi

49. Prothorax is large and elongated with shield like pronotum in

(*a*) Cockroaches (*b*) Grasshoppers
(*c*) Praying Mantids (*d*) All of these

50. Fore legs are raptorial

(*a*) Cockroaches (*b*) Grasshoppers
(*c*) Praying Mantids (*d*) All of these

51. Polymorphism is seen in the insect order

(*a*) Zoraptera (*b*) Dermaptera
(*c*) Isoptera (*d*) Plecoptera

52. Locust Warning Organization was established in

(*a*) 1939 (*b*) 1949
(*c*) 1959 (*d*) 1925

53. Author of Systema Naturae is

(*a*) Koenig (*b*) Linnaeus
(*c*) Anderson (*d*) Fabricious

54. Indian Museum was established in the year

(*a*) 1975 (*b*) 1875
(*c*) 1865 (*d*) 1965

55. Which insect order is most commonly found in soil litter?

(*a*) Collembola (*b*) Neuroptera
(*c*) Orthoptera (*d*) Lepidoptera

56. The Insecticide Act 1968 came into force in

(*a*) 1968 (*b*) 1966
(*c*) 1971 (*d*) 1975

57. Which insect order is most closely related to Dictyoptera?

(*a*) Orthoptera (*b*) Plecoptera
(*c*) Thysanura (*d*) Hymenoptera

58. What do the orders Mantodea, Dermaptera and Isoptera have in common?

(*a*) Winglessness (*b*) Herbivory
(*c*) Chewing mouthparts (*d*) All of these

59. Lobsters and shrimp belong to which class of arthropods?

(*a*) Crustacea (*b*) Arachnida
(*c*) Myripoda (*d*) Xiphosura

60. The order Hemiptera contains which of the following

(*a*) Bed bugs and stink bugs
(*b*) Chewing and sucking lice
(*c*) Roaches and mantids
(*d*) Crickets and grasshoppers

61. Which of the following is not holometabolous

(*a*) Siphonaptera
(*b*) Hymenoptera
(*c*) Thysanoptera
(*d*) Neuroptera

62. Which of the following order is exclusively herbivorous

(*a*) Trichoptera
(*b*) Odonata
(*c*) Thysanoptera
(*d*) Coleoptera

63. Stridulatory organs are present on the hind femur in

(*a*) Acridoidea
(*b*) Grylloidea
(*c*) Tettigonioidea
(*d*) All of these

64. Eggs are enclosed in ootheca in

(*a*) Orthoptera
(*b*) Ephemeroptera
(*c*) Dictyoptera
(*d*) All of these

65. All ametabolous insects are

(*a*) Predatory
(*b*) Wingless
(*c*) Endognathous
(*d*) All of these

66. Immatures of the order Neuroptera would be classified as

(*a*) Scavengers
(*b*) Parasites
(*c*) Herbivores
(*d*) Predators

67. Theory of evolution was given by

(*a*) Darwin
(*b*) Linnaeus
(*c*) Aristotle
(*d*) Mendel

68. Paired gonophore is present in

(*a*) Odonata
(*b*) Ephemeroptera
(*c*) Dermaptera
(*d*) All of these

69. A locust swarm usually consists of

(*a*) Short horned grasshoppers
(*b*) Long horned grasshoppers
(*c*) Gryllids
(*d*) All of these

70. Tympanal organs are present on the 1st abdominal segment in

(*a*) Acridoidea
(*b*) Grylloidea
(*c*) Tettigonioidea
(*d*) All of these

71. Adult mite possess__________pairs of legs.

(a) Two (b) Three

(c) Four (d) Five

72. Crab belong to the class

(a) Onychophora (b) Arachnida

(c) Myriapoda (d) Crustacea

73. In Diplopoda respiration takes place by means of

(a) Green gland (b) Malpighian tubules

(c) Book lungs (d) Tracheae

74. Scorpions are included in the class

(a) Onychophora (b) Arachnida

(c) Myriapoda (d) Crustacea

75. Cerci are unsymmetrical in males of

(a) Odonata (b) Embioptera

(c) Plecoptera (d) Dermaptera

76. The Gladiator belongs to the insect order

(a) Phasmida (b) Orthoptera

(c) Grylloblattoidea (d) Mantophasmatodea

77. Completely parasitic order is

(a) Lepidoptera (b) Neuroptera

(c) Strepsiptera (d) Hymenoptera

78. The crustaceans breathe by means of

(a) Green gland (b) Gills

(c) Book lungs (d) Tracheae

79. In chilopoda, the poison gland is found in the

(a) First pairs of legs (b) Second pairs of legs

(c) Third pairs of legs (d) None of these

80. In Diplopoda, thorax is divided into

(a) 3 segments (b) 4 segments

(c) 5 segments (d) 6 segments

81. Symphylians have

(a) Two pairs of antennae (b) One pair of antennae

(c) Three pairs of antennae (d) Antennae absent

82. Last instar nymph moults into subimago (Pelicle) is an important characteristic of the insect order

(*a*) Odonata (*b*) Ephemeroptera

(*c*) Plecoptera (*d*) All of these

83. Which of the following is characteristic feature of insect order Phasmida?

(*a*) Male genitalia is asymmetrical

(*b*) Presence of a pair of repugnatorial glands in prothorax

(*c*) Autotomy power of regeneration

(*d*) All of these

84. Cerci are modified into unjointed forceps in

(*a*) Odonata (*b*) Ephemeroptera

(*c*) Plecoptera (*d*) Dermaptera

85. The body is divided into cephalothorax and abdomen in

(*a*) Arachnida (*b*) Insecta

(*c*) Diplopoda (*d*) Crustacea

86. The antennae chelicerae are found in

(*a*) Crustacea (*b*) Arachnida

(*c*) Onychophora (*d*) Symphyla

87. Which of the following order having maximum numbers of predatory insects?

(*a*) Hymenoptera (*b*) Diptera

(*c*) Coleoptera (*d*) Lepidoptera

88. The insect having scaly wings comes under

(*a*) Hymenoptera (*b*) Diptera

(*c*) Coleoptera (*d*) Lepidoptera

89. Naids are larval form of which of the following order

(*a*) Diptera (*b*) Coleoptera

(*c*) Odonata (*d*) Lepidoptera

90. White fly belong to the order

(*a*) Hymenoptera (*b*) Hemiptera

(*c*) Coleoptera (*d*) Lepidoptera

91. Brachypterous form of wings found in the insect order

(*a*) Orthoptera (*b*) Hymenoptera

(*c*) Coleoptera (*d*) Lepidoptera

92. Halares are found in the insect order

(a) Orthoptera (b) Hymenoptera

(c) Diptera (d) Lepidoptera

93. Semilooper belongs to order

(a) Orthoptera (b) Hymenoptera

(c) Coleoptera (d) Lepidoptera

94. Asymmetrical cerci is present in

(a) Embioptera (b) Mallophaga

(c) Collembola (d) Plecoptera

95. Which of the following is characteristic feature of insect order Ephemeroptera?

(a) Nymphs are aquatic and long lived whereas adults are terrestrial and very short lived

(b) Presence of large number of instars (23 to 45)

(c) Last instar nymph moults into subimago

(d) All of these

96. In Odonata, which of the following is true

(a) Antennae is short and filiform

(b) Presence of pterostigma between costa and radius

(c) Capable of flying in forward and backward direction

(d) All of these

97. Immature stage of Coleoptera is

(a) Nymph (b) Caterpillar

(c) Grub (d) Maggot

98. Immature stage of Lepidoptera is

(a) Nymph (b) Caterpillar

(c) Grub (d) Maggot

99. Immature stage of Diptera is

(a) Nymph (b) Caterpillar

(c) Grub (d) Maggot

100. Immature stage of Hemiptera is

(a) Nymph (b) Caterpillar

(c) Grub (d) Maggot

101. Pollination of flowers by insects is known as

(*a*) Insectophily (*b*) Entophily

(*c*) Entomophily (*d*) None of these

102. The locust belongs to the family

(*a*) Termitidae (*b*) Bombycidae

(*c*) Locustidae (*d*) Accrididae

103. Hemelytra is a form of wings found in

(*a*) Heteroptera (*b*) Homoptera

(*c*) Lepidoptera (*d*) Orthoptera

104. Which of the following family having parasitic insects?

(*a*) Dermastidae (*b*) Braconidae

(*c*) Locustidae (*d*) Accrididae

105. Honey bees belongs to the order

(*a*) Lepidoptera (*b*) Diptera

(*c*) Hemiptera (*d*) None of these

106. In mite respiration takes place through

(*a*) Skin (*b*) Trachea

(*c*) Book Lungs (*d*) Lungs

107. Which of the following is the odd one?

(*a*) Housefly (*b*) Alderfly

(*c*) Cattlefly (*d*) House fly

108. Autotomy is observed in

(*a*) Collembola (*b*) Diptera

(*c*) Phasmida (*d*) Dictyoptera

109. Order containing single genus is

(*a*) Psocoptera (*b*) Mallophaga

(*c*) Collembola (*d*) Zoraptera

110. Number of sub-phylum found in Arthropoda

(*a*) 7 (*b*) 8

(*c*) 6 (*d*) 7

111. Example of the smallest order of class Insecta are

(*a*) Odonata (*b*) Coleoptera

(*c*) Lepidoptera (*d*) Zoraptera

112. Which of the following is an example of beautiful order of insects?

(a) Odonata (b) Coleoptera

(c) Lepidoptera (d) Zoraptera

113. Who was the first to study Indian insect type?

(a) Koening (b) Fabricius

(c) Lefroy (d) Linnaeus

114. Wingless arthropods with 6 abdominal segments are found in which order?

(a) Diplura (b) Thysanura

(c) Collembola (d) Protura

115. Bombay Natural History Society was founded in the year

(a) 1800 (b) 1883

(c) 1884 (d) 1885

116. The International Code of Zoological Nomenclature contain

(a) 58 articles (b) 87 articles

(c) 60 articles (d) 89 articles

117. Naids are aquatic in

(a) Plecoptera (b) Ephemeroptera

(c) Odonata (d) All of these

118. Zygoptera and Anisoptera are suborders of

(a) Thysanura (b) Diplura

(c) Odonata (d) Ephemeroptera

119. Which of the following is called Water Scorpion?

(a) Belostoma (b) Nepa

(c) Corixa (d) Notonecta

120. Millipedes belongs to the class

(a) Crustacea (b) Diplopoda

(c) Arachnida (d) Insecta

121. Anamorphosis is found in the order

(a) Protura (b) Orthoptera

(c) Dictyoptera (d) Coleoptera

122. Example of secondarily wingless insect is

(a) Head Louse (b) Housefly

(c) Honey bee (d) Beetles

123. Mite, ticks and spider comes under the class

(*a*) Insecta (*b*) Arachnida
(*c*) Pauropoda (*d*) Diplopoda

124. Primarily wingless insects are found in which order

(*a*) Hemiptera (*b*) Protura
(*c*) Odonata (*d*) Orthoptera

125. Most important character of Dictyoptera is

(*a*) Elytra (*b*) Hemelytra
(*c*) Large anal lobe in hind wing (*d*) None of these

126. Most dominant characteristics of insects are

(*a*) Antennae (*b*) Small size
(*c*) Strong flier (*d*) All of these

127. Snow fleas belongs to the order

(*a*) Protura (*b*) Siphonoptera
(*c*) Collembola (*d*) Mallophaga

128. Terminal 11th abdominal segment with a median telson is characteristic of

(*a*) Thysanura (*b*) Diplura
(*c*) Protura (*d*) Collembola

129. Which of the following is characteristic of insect order Odonata?

(*a*) Adults are terrestrial while nymphs are aquatic
(*b*) Both nymphs and adults are predatory
(*c*) Nymphs with prehensile labial mask and with rectal or caudal gills
(*d*) All of these

130. Branch of Science that deals with study of animal behavior is called

(*a*) Etiology (*b*) Ethology
(*c*) Entomology (*d*) Zoology

131. Light-producing substance in firefly is in

(*a*) Pyrethrin (*b*) BOAA
(*c*) Allethrin (*d*) Luciferin

132. Light producing insect family is

(*a*) Curculionidae (*b*) Chrysomelidae
(*c*) Lampyiridae (*d*) None of these

133. 1st imperial entomologist was

(a) HS Pruthi
(b) HM Lefroy
(c) Ronald Ross
(d) None of these

134. First Indian insect collector

(a) HS Pruthi
(b) HM Lefroy
(c) Ronald Ross
(d) Ramakrishna Ayyar

135. The word Insect is derived from

(a) English
(b) Latin
(c) French
(d) Greek

136. The word Arthropoda is derived from

(a) English
(b) Latin
(c) French
(d) Greek

137. Which of the following is a character of Paraneopteran Orders (Hemipteroid complex)?

(a) Sucking and Piercing mouth parts
(b) Absence of anal lobe
(c) Presence of minimum number of Malphigian tubules
(d) All of these

138. A pollen basket formed from long curved hairs on the hind leg of a bee is known as

(a) Corbicula
(b) Cordulia
(c) Colias
(d) None of these

139. All Apterygotes are

(a) Primitively winged
(b) Secondarily wingless
(c) Secondarily winged
(d) Primarily wingless

140. Family of grasshopper is

(a) Arctiidae
(b) Acrididae
(c) Pyralidae
(d) Noctuidae

141. Family of Red hairy caterpillar is

(a) Arctiidae
(b) Acrididae
(c) Pyralidae
(d) Noctuidae

142. Family of sugarcane leaf hopper is

(a) Pyralidae
(b) Acrididae
(c) Fulgoridae
(d) Noctuidae

143. The order of termite is

(*a*) Lepidoptera (*b*) Coleoptera
(*c*) Diptera (*d*) Isoptera

144. Systematists whose classification is based on the regency of common ancestry, ignoring similarities and differences between existing organisms are called

(*a*) Cladists (*b*) Pheneticists
(*c*) Phyleticists (*d*) None of these

145. Linker of phylum Annelida and Arthropoda is

(*a*) Diplopoda (*b*) Insecta
(*c*) Onychophora (*d*) Trilobita

146. Which one of the following characters did not fit with Crustaceans?

(*a*) Two pairs of antennae
(*b*) Three pairs of legs
(*c*) No wings present
(*d*) Body divided into Cephalothorax and abdomen

147. Which of the following characteristics are not found in insects?

(*a*) Segmented body (*b*) Closed circulation
(*c*) Jointed legs (*d*) Respiration through trachea

148. Fossil records indicate that insect evolved during

(*a*) Devonion period (*b*) Cambrian period
(*c*) Silurian period (*d*) Precambrian period

149. Which insects are more or less the same as fossil records?

(*a*) Moth (*b*) Cockroach
(*c*) Butterfly (*d*) Wasp

150. Branches of Biology which deals with rule, principle and basis is called

(*a*) Taxonomy (*b*) Classification
(*c*) Systemic (*d*) None of these

151. Naming and ordering of species on the basis of their characteristics is called

(*a*) Taxonomy (*b*) Classification
(*c*) Systemic (*d*) None of these

152. Zoological Survey of India (ZSI) is situated at

(*a*) New Delhi (*b*) Kolkata
(*c*) Mumbai (*d*) Hyderabad

153. Scientific name should be printed in

(*a*) Small letters (*b*) Capital letters

(*c*) Italic letters (*d*) Bold letters

154. Long-horned grasshoppers belongs to the sub order

(*a*) Ensifera (*b*) Caelifera

(*c*) Brachycera (*d*) Nematocera

155. In binomial nomenclature different classes possessing common characters are included in

(*a*) Sub class (*b*) Order

(*c*) Super class (*d*) Phylum

156. Phyla comprises triploblastic and acoelomate animals

(*a*) Coelenterate (*b*) Porifera

(*c*) Ctenophore (*d*) Annelida

157. Indian Board of Wild Life was established in

(*a*) 1952 (*b*) 1953

(*c*) 1954 (*d*) 1956

158. The Fauna of British India was started in

(*a*) 1814 (*b*) 1876

(*c*) 1888 (*d*) 1892

159. The Fauna of British India is written by

(*a*) Hampson (*b*) Lefroy

(*c*) Stebbing (*d*) Fletcher

160. The first president of Entomological Society of India was

(*a*) HS Pruthi (*b*) Niceville

(*c*) Afzal Hussain (*d*) Tashir Ahmad

161. Who proposed the broad classification of Animal Kingdom?

(*a*) Linnaeus (*b*) Aristotle

(*c*) Hymen (*d*) Huxley

162. Who was considered as the father of Biological Classification?

(*a*) Linnaeus (*b*) Aristotle

(*c*) Cuvier (*d*) Darwin

163. Who was considered as the father of Taxonomy?

(*a*) Linnaeus (*b*) Aristotle

(*c*) Mayr (*d*) Darwin

164. The first volume of "Fauna of British India" was written on which insects

(*a*) Aphids (*b*) Thrips

(*c*) Butterflies (*d*) Moths

165. Small apterous insects, compounds eyes are reduced or absent, ectoparasites of birds and mammals comes into

(*a*) Thysanura (*b*) Thysanoptera

(*c*) Mallophaga (*d*) Hymenoptera

166. Stick insects belongs to the order

(*a*) Orthoptera (*b*) Phasmida

(*c*) Dictyoptera (*d*) Embioptera

167. Number of insect species known at present

(*a*) About 50,000 (*b*) About 1 million

(*c*) About 1 billion (*d*) About 3 million

168. Which of the following has agricultural importance?

(*a*) *Hydra* spp. (*b*) *Culex* spp.

(*c*) *Hirudinea* spp. (*d*) *Pheritima* spp.

169. Carbon dating is useful when determining the age of fossil less than__________years old.

(*a*) 10,000 years (*b*) 15,000 years

(*c*) 60,000 years (*d*) 25,000 years

170. The first insect fossil was found in

(*a*) Scotland (*b*) England

(*c*) USA (*d*) Russia

171. Study of the type and diversity of organisms and the relationships amongst them is known as

(*a*) Taxonomy (*b*) Classification

(*c*) Systematics (*d*) Biodiversity

172. The theory and practice of identification, description, naming and classification of organisms is known as

(*a*) Systematics (*b*) Classification

(*c*) Taxonomy (*d*) Biodiversity

173. The group of real organisms recognized as a formal unit at any level of hierarchic classification is known as

(*a*) Taxon (*b*) Genus

(*c*) Family (*d*) Species

174. Which of the following is not a principal character of Phylum Arthropoda?

(*a*) Segmented body

(*b*) Paired segmented appendages

(*c*) Bilateral symmetry

(*d*) Closed circulatory system

175. The subphylum of Arthropoda which contain only marine fossils

(*a*) Trilobita (*b*) Chelicerata

(*c*) Crustacea (*d*) Uniramia

176. Four pairs of biramous appendages covered by carapace is characteristic feature of

(*a*) Trilobita (*b*) Chelicerata

(*c*) Crustacea (*d*) Uniramia

177. The group of inbreeding natural populations which are reproductively isolated from other such groups is known as

(*a*) Taxon (*b*) Genus

(*c*) Species (*d*) Family

178. The tabular arrangement of species, genera, orders or other classification categories according to characters and traits that serve to identify them is called as

(*a*) Ranking (*b*) Key

(*c*) Hierarchy (*d*) Tree

179. When one name has been given to two or different taxa is called as

(*a*) Homonym (*b*) Synonyms

(*c*) Autonym (*d*) Isonym

180. Different names given to the same taxon are called as

(*a*) Homonym (*b*) Synonyms

(*c*) Autonym (*d*) Isonym

181. International Code of Zoological Nomenclature is the system of rules and recommendations authorized by the

(*a*) International Conference of Zoology

(*b*) International Congress of Zoology

(*c*) Indian Congress of Zoology

(*d*) International Zoology Association

182. __________, 1758 is arbitrarily assigned as the starting point of Zoological Nomenclature.

(*a*) 1st February (*b*) 31st March

(*c*) 1st January (*d*) 15th January

183. The earliest known fossil insect is *Rhyniella praecursor*, belongs to the order__________from the Devonian period of Scotland, discovered by Hirst and Maulik in 1926.

(*a*) Lepidoptera (*b*) Dictyoptera

(*c*) Collembola (*d*) Protura

184. According to the International Code of Zoological Nomenclature, superfamily will have and ending with

(*a*) inae (*b*) oidea

(*c*) tera (*d*) deae

185. Insect containing laterally compressed body belong to

(*a*) Collembola (*b*) Zoraptera

(*c*) Mallophaga (*d*) Siphonaptera

186. Which of the following is an Exopterygote holometabolus insect order?

(*a*) Thysanoptera (*b*) Hymenoptera

(*c*) Mecoptera (*d*) Diptera

187. Which of the following is a character of Polyneoptera (Orthopteroid Group)?

(*a*) Mandibulate mouth parts

(*b*) Hind wings bearing a large anal lobe

(*c*) Presence of cerci

(*d*) All of these

188. According to the International Code of Zoological Nomenclature, subfamily will have an ending with

(*a*) inae (*b*) oidea

(*c*) tera (*d*) deae

189. The arrangement of organisms into groups on the basis of their relationship is called as

(*a*) Systematics (*b*) Classification

(*c*) Taxonomy (*d*) Biodiversity

190. The level of taxonomy concerned with characterization and naming of a species is

(*a*) Alpha taxonomy (*b*) Beta taxonomy

(*c*) Gamma taxonomy (*d*) None of these

191. The level of taxonomy concerned with the arrangement of species in the biological hierarchy is

(*a*) Alpha taxonomy
(*b*) Beta taxonomy
(*c*) Gamma taxonomy
(*d*) None of these

192. The level of taxonomy concerned with the studies on intra-specific populations to understand speciation, evolution rates and trends is

(*a*) Alpha taxonomy
(*b*) Beta taxonomy
(*c*) Gamma taxonomy
(*d*) None of these

193. Local population of a species or the community of potential interbreeding individual at a given locality is called as

(*a*) Taxa
(*b*) Race
(*c*) Deme
(*d*) None of these

194. An intermediate population unit which exists between species and deme is called

(*a*) Taxa
(*b*) Biotype
(*c*) Race
(*d*) None of these

195. Members of Mantophasmatodea have superficial resemblance to

(*a*) Dictypotera
(*b*) Phasmida
(*c*) Thysanura
(*d*) Diplura

196. Diplura, Protura and Collembola have__________mouth parts

(*a*) Ectognathus
(*b*) Endognathus
(*c*) Both (*a*) & (*b*)
(*d*) None of these

197. Thysanura and Pterygota have__________mouth parts

(*a*) Ectognathus
(*b*) Endognathus
(*c*) Both (*a*) & (*b*)
(*d*) None of these

198. In Odonata secondary copulatory apparatus is present on__________sterna of adult males

(*a*) 1st and 4th
(*b*) 2nd and 3rd
(*c*) 4th and 5th
(*d*) 6th and 7th

199. Highly evolved insect order is

(*a*) Diptera
(*b*) Coleoptera
(*c*) Hemiptera
(*d*) Hymenoptera

200. Structurally more specialized order is

(*a*) Diptera
(*b*) Coleoptera
(*c*) Lepidoptera
(*d*) Hymenoptera

201. Which of the following orders are absent in India?

(*a*) Zoraptera and Grylloblattoidea (*b*) Zoraptera and Embioptera

(*c*) Zoraptera and Collembola (*d*) Zoraptera and Trichoptera

202. Permothemis (Archodonate fossil) is a connecting link between Ephemeroptera and__________.

(*a*) Collembola (*b*) Phasmida

(*c*) Odonata (*d*) Zoraptera

203. Tympanal organs on the fore tibia and stridulartory organs on tegmina are present in

(*a*) Long horned grasshoppers (*b*) Gryllids

(*c*) Cockroach (*d*) Both (*a*) & (*b*)

204. In cockroaches, ocelli are reduced and are represented by

(*a*) Eye (*b*) Compound eye

(*c*) Fenestrae (*d*) Ogle

205. In which of the families of Isoptera, workers are absent?

(*a*) Mastotermitidae (*b*) Kalotermitidae

(*c*) Termopsidae (*d*) All of these

206. Other insects and arthropods living in the termitarium are collectively known as

(*a*) Symphilans (*b*) Termitophiles

(*c*) Co-type (*d*) None of these

207. Connecting link between the orthopteroid and hemipteroid groups is

(*a*) Isoptera (*b*) Zoraptera

(*c*) Ephimeroptera (*d*) Hemiptera

208. Presence of pearmans organ on coxa which act as stridulatory organs is seen in

(*a*) Dictyoptera (*b*) Psocoptera

(*c*) Odonata (*d*) Ephimeroptera

209. In Thysanoptera, females bear a saw like serrate ovipositor in the suborder

(*a*) Terebrantia (*b*) Tubulifera

(*c*) Both (*a*) & (*b*) (*d*) None of these

210. Hind tibia with an apical large movable spur is a characteristic features of__________family.

(*a*) Aphididae (*b*) Cicadellidae

(*c*) Delphacidae (*d*) Agromyzidae

211. Silver fish belongs to the order

(a) Thysanura
(b) Diplura
(c) Protura
(d) Collembola

212. Antennae are absent in

(a) Dermaptera
(b) Mallophaga
(c) Protura
(d) Collembola

213. In aphids cornicles are present on__________abdominal segment.

(a) 2nd
(b) 3rd
(c) 5th
(d) 7th

214. Triungulin larvae is seen in the family

(a) Chrysomelidae
(b) Meloidae
(c) Coccinellidae
(d) Agromyzidae

215. In weevils an elongated rostrum is formed due to prolongation of vertex and

(a) Occiput
(b) Frons
(c) Genae
(d) Clypeus

216. In Lepidoptera, mouthparts are well developed with the functional mandibles in the group

(a) Acanthopteroctetidae
(b) Eriocraniidae
(c) Micropterigidae
(d) Tortricidae

217. Infection caused by dipteran larvae in the external body cavities of man and domestic animals is termed as

(a) Myiasis
(b) Mycosis
(c) Antibiosis
(d) None of these

218. Vein less wing bearing hymenopteran parasitoids belongs to the family

(a) Trichogramatidae
(b) Braconidae
(c) Ichnumunidae
(d) Platygasteridae

219. Which of the following is the basic unit of classification?

(a) Species
(b) Genus
(c) Family
(d) Order

220. Cohort is present in between

(a) Class and Order
(b) Family and Genus
(c) Phylum and Class
(d) Order and Family

221. The connecting link in between Apterygota and Pterygota is

(*a*) Thysanura (*b*) Diplura

(*c*) Protura (*d*) Collembola

222. Sub-species is included in which type of classification

(*a*) Binomial classification (*b*) Trinomial classification

(*c*) Natural classification (*d*) Artificial classification

223. Insects are classified based on

(*a*) Antennae (*b*) Wings

(*c*) Legs (*d*) Mouth parts

224. A group of organisms at any particular level in a classification system is called a

(*a*) Species (*b*) Genus

(*c*) Taxon (*d*) Phylum

225. Scientific names higher than genus are

(*a*) Capitalized but not printed distinctively, italicized, or underlined

(*b*) Italicized or underlined

(*c*) Capitalized and italicized or underlined

(*d*) Italicized or underlined with the first word capitalized

226. As a practical definition of species, one can say that they are a group of organisms characterized by all of the following except they

(*a*) Do not normally interbreed with other species in nature

(*b*) Can be distinguished from other species

(*c*) Are incapable of hybridization with other species

(*d*) Remain relatively constant

227. The institution of the classification level of domain was prompted by

(*a*) The overabundance of kingdoms

(*b*) Cladistics

(*c*) Great differences among bacteria

(*d*) The taxonomic problem presented by viruses

228. Viruses are denied a kingdom of their own because

(*a*) They are too poorly understood (*b*) They are too small

(*c*) Their genetics cannot be determined (*d*) They are not organisms

229. Viruses are assigned to the kingdom

(*a*) Archaebacteria (*b*) Protista

(*c*) Fungi (*d*) None of the above

230. A major problem with polynomial names was that they were

(*a*) Inaccurate (*b*) Cumbersome

(*c*) In Greek (*d*) In Latin

231. Expansion of ICZN is

(*a*) International Code of Zoological Nomenclature

(*b*) International Conference on Zoological Nomenclature

(*c*) International Commission on Zoological Nomenclature

(*d*) Indian Council on Zoological Nomenclature

232. The order Dictyoptera is divided into two sub orders namely

(*a*) Blattidae and Mantdae (*b*) Corrodentia and Anoplura

(*c*) Blattaria and Mantodea (*d*) None of these

233. The plant alternation of generations is characterized by__________meiosis.

(*a*) Zygotic (*b*) Somatic

(*c*) Sporic (*d*) Generic

234. When writing species epithets, it is sometimes permissible to abbreviate the__________name.

(*a*) Species (*b*) Phylum

(*c*) Order (*d*) Genus

235. True multi-cellularity occurs only in

(*a*) Animals (*b*) Eukaryotes

(*c*) Plants, animals and protists (*d*) Plants, animals and fungi

236. A related group of genera comprise

(*a*) An order (*b*) A family

(*c*) A class (*d*) A division

237. Prokaryotic organisms make up the

(*a*) Archaebacteria, Eubacteria, and Protists

(*b*) Archaebacteria and Protists

(*c*) Protists and Eubacteria

(*d*) Eubacteria and Archaebacteria

238. In the current taxonomic system, families are grouped into

(*a*) Classes (*b*) Phyla

(*c*) Orders (*d*) Divisions

239. Of the six kingdoms now recognized

(*a*) Two are plants and four are animals

(*b*) Four are eukaryotes and two are prokaryotes

(*c*) Four are macroscopic and two are microscopic

(*d*) Two are eukaryotes and four are prokaryotes

240. In plant taxonomy, a division is the equivalent to__________.

(*a*) Family (*b*) Order

(*c*) Class (*d*) Phylum

241. Unicellular eukaryotes are grouped into

(*a*) Fungi and protists

(*b*) Fungi and Eubacteria

(*c*) Only protists

(*d*) Fungi, protists, and Eubacteria

242. In printed scientific names, only the__________is capitalized.

(*a*) Family (*b*) Class

(*c*) Kingdom (*d*) Genus

243. In the early 1700's, the classification system used was

(*a*) Polynomial (*b*) Pentanomial

(*c*) Binomial (*d*) Tetranomia

244. In printing scientific names, it is conventional for only the__________to be underlined or italicized.

(*a*) Binomials (*b*) Genus

(*c*) Species (*d*) Division

245. Characteristics that have arisen as a result of common evolutionary descent are said to be

(*a*) Analogous (*b*) Homogenous

(*c*) Heterogamous (*d*) Homologous

246. The number of categories recognized by the taxonomists is

(*a*) 18 (*b*) 19

(*c*) 17 (*d*) 16

247. Insect are the most abundant and diversified of all animals due to their

(a) Segmented body (b) Antennae
(c) Three pairs of legs (d) Capacity of flight

248. A kingdom that also comprises a domain is

(a) Eubacteria (b) Protista
(c) Fungi (d) Prokaryota

249. The oldest level of taxonomic classification is

(a) Class (b) Family
(c) Genus (d) Phylum

250. Multicellular organisms and unicellular yeasts are contained in

(a) Animalia (b) Plantae
(c) Protista (d) Fungi

251. A single specimen selected by the author of a species as its type, or the only specimen known at the time of description is known as

(a) Holotype (b) Allotype
(c) Syntype (d) Paratype

252. A term designating a specimen of opposite sex to the holotype is

(a) Allotype (b) Lectotype
(c) Neotype (d) Topotype

253. Throughout the world, official species names are in

(a) Latin (b) Greek
(c) English (d) French

254. The arthropods which lack antennae and typically have six pairs of appendages belongs to

(a) Trilobita (b) Chelicerata
(c) Crustacea (d) Uniramia

255. The respiratory organs in Arachnida is

(a) Book lungs (b) Trachea
(c) Cuticle (d) Green gland

256. The arthropods which posses two pairs of antennae

(a) Trilobita (b) Chelicerata
(c) Crustacea (d) Uniramia

257. A taxon consisting of the most closely related species is called a(n)_________.

(*a*) Family (*b*) Order
(*c*) Genus (*d*) Phylum

258. The binomial system of classification was developed by

(*a*) Darwin (*b*) Wallace
(*c*) Linnaeus (*d*) Malthus

259. Which one of the following sequences shows the correct hierarchy of classification, going from the most inclusive to the least inclusive?

(*a*) Kingdom, Domain, Phylum, Order, Class, Family, Genus, Species
(*b*) Domain, Kingdom, Phylum, Class, Order, Family, Genus, Species
(*c*) Genus, Species, Kingdom, Phylum, Order, Class, Family
(*d*) Domain, Phylum, Kingdom, Genus, Species, Family, Order, Class

260. The separation of the Archaebacteria into a separate domain indicates

(*a*) That these organisms are very different from eubacteria and from eukaryotes
(*b*) That organisms should never be classified according to how they appear (morphological characteristics)
(*c*) That early taxonomists rushed to classify them without closely examining them
(*d*) That optical viewing techniques have greatly improved so that we can better view these microorganisms

261. Which of the following is not a principle character of Crustacea?

(*a*) 5 pairs of leg (*b*) 2 pairs of antennae
(*c*) Respiration by book lungs (*d*) Body covered by carapace

262. Centipedes have

(*a*) One pair of legs on every segment (*b*) Two pairs of legs on every segment
(*c*) No legs (*d*) 5 antennae

Fill in the Blanks

1. Systematic includes taxonomy and_________.

2. To classify biological organisms, hierarchal concept of classification was introduced by_________.

3. In trinomial nomenclature, the lowermost category for classifying an organism is_________.

4. A_________consists of a large number of inter communicating gene pool.

5. When a key is constructed based on a single character it is known as__________.

6. When a key is constructed based on many characters, it is known as__________.

7. The key is constructed based on the evolutionary relationship, here the group appears only once is called as__________.

8. In apterygota mandibular articulation in head is__________.

9. In pterygota mandibular articulation in head is__________.

10. In exopterygota wings are developed__________.

11. In endopterygota wings are developed__________.

12. Panarpoid insects are__________and they have larval and pupal stage.

13. The study of kinds and diversity of organisms and the evolutionary relationships among them is called__________.

14. __________gave the modern classification system.

15. The assignment of a distinctive name to each species is called__________.

16. The constant mutation rate of organisms is called__________clock.

17. The members of kingdom__________are prokaryotes.

18. The groups showing similarities due to single ancestors are called__________groups.

19. The systematic in which relationships are developed by study of fossils of ancestors of closely related animals is called__________systematics.

20. The resemblances that result from organisms adapting under similar evolutionary pressures are called__________.

21. The founders of__________taxonomy believed that there are no criteria for grouping taxa.

22. Characters that have arisen since common ancestry with the out-group are called derived characters or__________.

23. The absence of a central point or axis around which body parts are equally distributed is called__________.

24. __________is a fluid-filled space in which the internal organs can be suspended and separated from the body wall.

Match the Following

1. Match the insets with their order

(a) Blattodea	i) Earwigs
(b) Coleoptera	ii) Springtails
(c) Collembola	iii) Beetles
(d) Dermaptera	iv) Cockroaches

2. Match the insets with their order

(a) Diptera	i) Rock crawlers
(b) Embioptera	ii) Mayflies
(c) Ephemeroptera	iii) Flies
(d) Grylloblattodea	iv) Web spinners

3. Match the insets with their order

(a) Hemiptera	i) Termites
(b) Hymenoptera	ii) Butterflies and moths
(c) Isoptera	iii) Bugs, cicada, hoppers and scale insects
(d) Lepidoptera	iv) Ants, bees, wasps and sawflies

4. Match the insets with their order

(a) Mantodea	i) Lacewings, alderflies, snake-flies and antlions
(b) Mallophaga	ii) Biting lice and bird lice
(c) Mecoptera	iii) Praying mantis
(d) Neuroptera	iv) Scorpion flies

5. Match the insets with their order

(a) Odonata	i) Stoneflies
(b) Orthoptera	ii) Stick and leaf insects
(c) Phasmatodea	iii) Grasshoppers, locusts and crickets
(d) Plecoptera	iv) Dragonflies and damselflies

6. Match the insets with their order

(a) Protura	i) Proturans
(b) Psocoptera	ii) Fleas
(c) Siphonaptera	iii) Sucking lice
(d) Siphunculata	iv) Booklice

7. **Match the insets with their order**

(a) Strepsiptera	i) Bristletails and silverfish
(b) Thysanoptera	ii) Thrips
(c) Thysanura	iii) Caddis flies
(d) Trichoptera	iv) Stylopids

8. **Match the insets with their family**

(a) Carabidae	i) Fireflies
(b) Coccinellidae	ii) Dung Beetles
(c) Lampyridae	iii) Ground Beetles
(d) Scarabaeidae	iv) Ladybird Beetles

9. **Match the insets with their family**

(a) Culicidae	i) Mosquitoes
(b) Tabanidae	ii) House Flies
(c) Tephritidae	iii) Fruit Flies
(d) Muscidae	iv) Horse Flies

10. **Match the arthropods with their class**

(a) Crustaceans	i) Crabs
(b) Arachnids	ii) Millipedes
(c) Myriapods	iii) Spiders
(d) Insects	iv) Cricket

11. **Match the class of arthropods with their number of body segments**

(a) Crustaceans	i) 2
(b) Arachnids	ii) 2 - 3
(c) Centipedes	iii) 3
(d) Insects	iv) 100 or more

12. **Match the class of arthropods with their number of legs**

(a) Crustaceans	i) 100 or more pairs
(b) Arachnids	ii) 5 or more pairs
(c) Centipedes	iii) 3 pairs
(d) Insects	iv) 4 pairs

13. **Match the insects with modification of their cerci**

(a) Orthoptera	i) Un-segmented
(b) Embioptera	ii) Unsymmetrical in males
(c) Dermaptera	iii) Many segmented
(d) Dictyoptera	iv) Forceps like

14. Match the following insects with their families

(a)	Plant hoppers	i)	Delphacidae
(b)	Frog hoppers	ii)	Membracidae
(c)	Tree hoppers	iii)	Cercopidae
(d)	Leaf hoppers	iv)	Cicadellidae

15. Match the following insects with their families

(a)	Ground beetles	i)	Bruchidae
(b)	Tiger beetles	ii)	Cicindelidae
(c)	Click beetles	iii)	Elateridae
(d)	Seed beetles	iv)	Carabidae

Answers

Multiple Choice Questions

1	(b)	21	(d)	41	(a)	61	(c)
2	(d)	22	(b)	42	(d)	62	(c)
3	(d)	23	(c)	43	(c)	63	(a)
4	(c)	24	(b)	44	(a)	64	(c)
5	(b)	25	(b)	45	(d)	65	(b)
6	(b)	26	(a)	46	(c)	66	(d)
7	(a)	27	(a)	47	(b)	67	(a)
8	(c)	28	(a)	48	(a)	68	(b)
9	(b)	29	(c)	49	(a)	69	(a)
10	(b)	30	(a)	50	(c)	70	(a)
11	(a)	31	(b)	51	(c)	71	(c)
12	(d)	32	(d)	52	(a)	72	(d)
13	(d)	33	(c)	53	(b)	73	(d)
14	(d)	34	(d)	54	(b)	74	(b)
15	(a)	35	(a)	55	(a)	75	(b)
16	(c)	36	(c)	56	(c)	76	(d)
17	(d)	37	(b)	57	(a)	77	(c)
18	(a)	38	(a)	58	(d)	78	(b)
19	(b)	39	(d)	59	(a)	79	(a)
20	(c)	40	(a)	60	(a)	80	(a)

81	(b)	115	(b)	149	(b)	183	(c)
82	(b)	116	(b)	150	(b)	184	(b)
83	(d)	117	(d)	151	(a)	185	(d)
84	(d)	118	(c)	152	(b)	186	(a)
85	(d)	119	(b)	153	(c)	187	(d)
86	(b)	120	(b)	154	(d)	188	(a)
87	(c)	121	(a)	155	(a)	189	(b)
88	(d)	122	(a)	156	(d)	190	(a)
89	(c)	123	(b)	157	(a)	191	(b)
90	(b)	124	(b)	158	(d)	192	(c)
91	(a)	125	(c)	159	(a)	193	(c)
92	(c)	126	(d)	160	(c)	194	(c)
93	(d)	127	(c)	161	(c)	195	(b)
94	(a)	128	(c)	162	(b)	196	(b)
95	(d)	129	(d)	163	(a)	197	(a)
96	(d)	130	(b)	164	(d)	198	(b)
97	(c)	131	(d)	165	(c)	199	(d)
98	(b)	132	(c)	166	(b)	200	(a)
99	(d)	133	(b)	167	(b)	201	(a)
100	(a)	134	(d)	168	(d)	202	(c)
101	(c)	135	(b)	169	(c)	203	(d)
102	(d)	136	(d)	170	(a)	204	(c)
103	(a)	137	(d)	171	(c)	205	(d)
104	(b)	138	(a)	172	(c)	206	(b)
105	(d)	139	(d)	173	(a)	207	(b)
106	(c)	140	(b)	174	(d)	208	(b)
107	(a)	141	(d)	175	(a)	209	(a)
108	(c)	142	(c)	176	(a)	210	(c)
109	(d)	143	(d)	177	(c)	211	(a)
110	(b)	144	(c)	178	(b)	212	(c)
111	(d)	145	(c)	179	(a)	213	(c)
112	(c)	146	(b)	180	(b)	214	(b)
113	(b)	147	(b)	181	(b)	215	(b)
114	(c)	148	(a)	182	(c)	216	(c)

217	(a)	229	(d)	241	(c)	253	(a)
218	(d)	230	(b)	242	(d)	254	(b)
219	(a)	231	(a)	243	(a)	255	(a)
220	(a)	232	(c)	244	(a)	256	(c)
221	(a)	233	(c)	245	(d)	257	(c)
222	(b)	234	(d)	246	(a)	258	(c)
223	(b)	235	(d)	247	(d)	259	(b)
224	(c)	236	(b)	248	(a)	260	(a)
225	(a)	237	(d)	249	(c)	261	(d)
226	(c)	238	(c)	250	(d)	262	(a)
227	(c)	239	(b)	251	(a)		
228	(d)	240	(b)	252	(a)		

Fill in the Blanks

1. Evolution
2. Linnaeus (1758)
3. Subspecies
4. Species
5. Monothetic key
6. Polythetic key
7. Phylogenetic key
8. Monocondylic
9. Dicondylic
10. Externally
11. Internally
12. Holometabolous
13. Systematics
14. Karl von Linnaeus
15. Nomenclature
16. Molecular
17. Monera
18. Monophyletic

19. **Evolutionary**
20. **Analogies**
21. **Numerical**
22. **Synapomorphies**
23. **Asymmetry**
24. **Coelom**

Matching

1. **a - iv, b - iii, c - ii, d - i**
2. **a - iii, b - iv, c - ii, d - i**
3. **a - iii, b - iv, c - i, d - ii**
4. **a - iii, b - ii, c - iv, d - i**
5. **a - iv, b - iii, c - ii, d - i**
6. **a - i, b - iv, c - ii, d - iii**
7. **a - iv, b - ii, c - i, d - iii**
8. **a - iii, b - iv, c - i, d - ii**
9. **a - i, b - iv, c - iii, d - ii**
10. **a - i, b - iii, c - ii, d - iv**
11. **a - ii, b - i, c - iv, d - iii**
12. **a - ii, b - iv, c - i, d - iii**
13. **a - i, b - ii, c - iv, d - iii**
14. **a - i, b - iii, c - ii, d - iv**
15. **a - iv, b - ii, c - iii, d - i**

Chapter 2
Morphology

Multiple Choice Questions

1. **The upper most layer of the insect body is known as**
   (a) Cuticle (b) Body wall or integument
   (c) Skeleton (d) None of these

2. **Body wall of insect serves as an**
   (a) Exoskeleton (b) Inner skeleton
   (c) Both (a) & (b) (d) None of these

3. **Body wall of insect is composed of**
   (a) The cuticle (b) The epidermis or hypodermis
   (c) The basement membrane (d) All of these

4. **The cuticle is an outer most layer of an insect body which is a**
   (a) Non-cellular layer (b) Cellular layer
   (c) Simple layer (d) None of these

5. **Example of an exopteregote insect order with a pupal instar in its life cycle**
   (a) Lepidoptera (b) Heteroptera
   (c) Trichoptera (d) Thysanoptera

6. **Pterostigma is identification character of the insect order**
   (a) Lepidoptera (b) Phasmida
   (c) Odonata (d) Dermaptera

7. The cuticle is an outer most layer of an insect body which is a

(*a*) Complex structure (*b*) Non-cellular layer

(*c*) Mostly secreted by the hypodermis (*d*) All of these

8. The cuticle layer comprising of sub-layers

(*a*) Epicuticle (*b*) Exocuticle

(*c*) Endocuticle (*d*) All of these

9. The cuticle serves a lining to the fore and hind gut and tracheae which are originated from

(*a*) Ectoderm (*b*) Mesoderm

(*c*) Endoderm (*d*) All of these

10. The epicuticle of insect is the outer most layer, which is very thin about

(*a*) $< 4\ \mu$ (*b*) $> 4\ \mu$

(*c*) $< 2\ \mu$ (*d*) $< 1\ \mu$

11. The epicuticle of insect is the outer most layer, consists of hardened protein called

(*a*) Curcumin (*b*) Cuticulin

(*c*) Albumin (*d*) Prolin

12. Pore canals present in the exocuticle helps in the deposition of

(*a*) Endocuticle (*b*) Epicuticle

(*c*) Exocuticle (*d*) None of these

13. Epicuticle is differentiated in to which of the layers

(*a*) Cement layer and wax layer (*b*) Cuticulin

(*c*) Outer epicuticle and inner epicuticle (*d*) All of these

14. Exocuticle is the outer layer, much thicker with the composition of

(*a*) Chitin (*b*) Sclerotin

(*c*) Both (*a*) & (*b*) (*d*) None of these

15. Sclerotin is a__________tanned by phenolic substances, a hard, brown material which gives rigidity to cuticle.

(*a*) Sugar (*b*) Fat

(*c*) Protein (*d*) Vitamin

16. Endocuticle and exocuticle put together are called

(*a*) Epicuticle (*b*) Procuticle

(*c*) Epiproct (*d*) None of these

17. Endocuticle is colourless, soft, flexible and the inner most thickest layer of cuticle which contains

(a) Cuticulin (b) Chitin

(c) Arthropodin (d) Both (b) & (c)

18. Functions of the insect cuticle are

(a) Give proper shape of the body

(b) Protect internal organs and check water loss

(c) Formation of different internal organs like tentorium, apodeme etc.

(d) All of these

19. The epidermis or hypodermis is the middle layer of the body wall and forms a__________continuous layer of cells.

(a) Single (b) Double

(c) Triple (d) Few

20. Functions of epidermis are

(a) Cuticle secretion and produce moulting fluid

(b) Digestion and absorption of old cuticle

(c) Wound repairing and formation of epidermal glands

(d) All of these

21. Which of the following glands are originating from epidermis?

(a) Salivary glands, poisonous glands, wax glands and mucous glands

(b) Silk glands, excretory glands and moulting fluid glands

(c) Scent glands, lac glands and repellent glands

(d) All of these

22. The basement membrane is the inner layer of body wall, where hypodermal cells are attached, this is a__________thin membrane.

(a) Multi cellular (b) Acellular

(c) Single cell (d) None of these

23. The main constituent of cuticle is chitin, which is nitrogenous polysaccharide and polymer of

(a) N - acetylsucrosamine (b) M - acetylglucosamine

(c) N - acetylfructosamine (d) N - acetylglucosamine

24. Chitin is water in soluble but soluble in

(a) Dilute acids (b) Alkalies

(c) Organic solvents (d) All of these

25. Arthropodin is an untanned protein, which is

(*a*) Water soluble (*b*) Water insoluble

(*c*) Water stable (*d*) None of these

26. Sclerotin is a tanned protein, which is

(*a*) Water insoluble (*b*) Water soluble

(*c*) Water stable (*d*) None of these

27. Anamorphosis is encountered the following insect order

(*a*) Protura (*b*) Collembola

(*c*) Thysanura (*d*) Diplura

28. The first longitudinal wing vein is

(*a*) Costal vein (*b*) Median vein

(*c*) Anal vein (*d*) Humeral vein

29. Resilin is an elastic protein responsible for the flexibility of

(*a*) Leg sclerites (*b*) Wing sclerites

(*c*) Mandibular sclerites (*d*) All of these

30. The body of grasshopper is covered with hard plates, which are collectively known as

(*a*) Tergum (*b*) Sternum

(*c*) Pluron (*d*) Sclerites

31. The sclerites are attached to each other with a thin membrane known as grooves and the joints are called as

(*a*) Tergum (*b*) Sternum

(*c*) Pluron (*d*) Sutures

32. Cuticular ingrowths of body wall providing space for muscle attachment is known as

(*a*) Exoskeleton (*b*) Endoskeleton

(*c*) Exocuticle (*d*) None of these

33. An internal projection or hollow invagination of exoskeleton, provides internal areas for muscular attachment is known as

(*a*) Apodeme (*b*) Apophysis

(*c*) Apolysis (*d*) Spur

34. Solid projection of exoskeleton either internally or externally is known as

(*a*) Apodeme (*b*) Apophysis

(*c*) Apolysis (*d*) Spur

35. Cellular cuticular appendages have epidermal association and may be

(*a*) Unicellular (*b*) Multicellular

(*c*) Both (*a*) & (*b*) (*d*) None of these

36. Multicellular cuticular structure are

(*a*) Spur (*b*) Spine

(*c*) Both (*a*) & (*b*) (*d*) None of these

37. Multicellular cuticular structure Spur are

(*a*) Movable (*b*) Immovable

(*c*) Stable (*d*) None of these

38. Multicellular cuticular structure Spine are

(*a*) Movable (*b*) Immovable

(*c*) Stable (*d*) None of these

39. Unicellular cuticular structures clothing hairs and plumose hairs are found in

(*a*) Honey bee (*b*) Flies

(*c*) Moths and butterflies (*d*) Caterpillar

40. Unicellular cuticular structures bristles are found in

(*a*) Honey bee (*b*) Flies

(*c*) Moths and butterflies (*d*) Caterpillar

41. Unicellular cuticular structures scales (flattened out growth of body wall) are found in

(*a*) Honey bee (*b*) Flies

(*c*) Moths and butterflies (*d*) Caterpillar

42. Unicellular cuticular structures glandular seta are found in

(*a*) Honey bee (*b*) Flies

(*c*) Moths and butterflies (*d*) Caterpillar

43. Epidermal cell generating seta is known as

(*a*) Trichogen (*b*) Tormogen

(*c*) Tentorium (*d*) None of these

44. The socket forming cell housing trichogen is known as

(*a*) Trichogen (*b*) Tormogen

(*c*) Tentorium (*d*) None of these

45. "Chaetotaxy" is the study of arrangement of

(*a*) Trichogen (*b*) Tormogen

(*c*) Tentorium (*d*) Seta

46. The endoskeleton of head is known as

(*a*) Trichogen (*b*) Tormogen

(*c*) Tentorium (*d*) None of these

47. Head capsule of insect is formed by the fusion of how many hardened plates or sclerites

(*a*) Two (*b*) Six

(*c*) Five (*d*) Four

48. Head capsule of insect is formed by the fusion of many hardened plates or sclerites namely

(*a*) Ocellary and antennal (*b*) Intercalary and mandibular

(*c*) Maxillary and labial (*d*) All of these

49. Small hooks used to couple together fore and hindwings this type of wing coupling mechanism is known as

(*a*) Jugate (*b*) Hamulate

(*c*) Frenate (*d*) None of these

50. The type of wing coupling mechanism present in Hymenoptera

(*a*) Jugate (*b*) Hamulate

(*c*) Frenate (*d*) None of these

51. The sclerotized capsule without appendages is also known as

(*a*) Trichogen (*b*) Tormogen

(*c*) Tentorium (*d*) Cranium

52. A median bristle between the two claws on the feet of Diptera is known as__________.

(*a*) Tormogen (*b*) Tentorium

(*c*) Sutures (*d*) Empodium

53. The long axis of the head is vertical and mouth parts are ventral and the median line of the head forms right angle with the median line of the body in case of

(*a*) Hypognathus head (*b*) Prognathus head

(*c*) Opisthognathus head (*d*) None of these

54. The long axis of the head is horizontal and mouth parts are anterior in position, so the median line of the head is paralled with the median line of the body in case of

(*a*) Hypognathus head (*b*) Prognathus head

(*c*) Opisthognathus head (*d*) None of these

55. The head is directed backwards and mouth parts are posterioventral in case of

(a) Hypognathus head
(b) Prognathus head
(c) Opisthognathus head
(d) None of these

56. Among the sutures of insect head which is not a true suture

(a) Coronal suture
(b) Epistomal suture
(c) Occipital suture
(d) Frontal suture or epicranial suture

57. The part of the insect head above the frontal suture is

(a) Vertex
(b) Occiput
(c) Genae
(d) Clypeus

58. Ocelli and antennae are situated on

(a) Vertex region
(b) Occiput region
(c) Genae region
(d) None of these

59. The frontal cuticular plate of an insect head lies below the vertex extended to the clypeus and laterally to the genae is known as

(a) Frons or front
(b) Pronotum
(c) Occiput
(d) None of these

60. The lateral cuticular plates of an insect's head below and behind the eye is known as

(a) Vertex
(b) Occiput
(c) Genae
(d) Clypeus

61. A cuticular plate on the front of an insect's head immediately anterior to the frons and above the labrum is known as

(a) Vertex
(b) Occiput
(c) Genae
(d) Clypeus

62. Occiput is situated in the back portion of the head and lies between the vertex and neck, which is divided from the vertex and genae by

(a) Coronal suture
(b) Epistomal suture
(c) Occipital suture
(d) Frontal suture

63. The head appendages of insects are

(a) Ocelli and Compound eyes
(b) Antennae and Mouth parts
(c) Both (a) & (b)
(d) None of these

64. Compound eyes of insects are paired structures, are situated in which position of head capsule

(*a*) Dorsal (*b*) Lateral

(*c*) Dorsolateral (*d*) Ventral

65. Elongated sensory feelers located on the head of an insect bearing minute organs of touch and smell

(*a*) Cerci (*b*) Antennae

(*c*) Setae (*d*) Flagella

66. The basal segment of an antenna, by which it is attached with the head is known as

(*a*) Coxa (*b*) Humerous

(*c*) Scape (*d*) Pedicel

67. The second segment of an antenna which bears sensory apparatus, the organ of Johnston is called as

(*a*) Coxa (*b*) Scape

(*c*) Pedicel (*d*) None of these

68. The front or upper lip of an insect's mouth parts is known as

(*a*) Labrum (*b*) Mandibles

(*c*) Maxillae (*d*) Labium

69. The lower lip of an insect mouth parts is known as

(*a*) Labrum (*b*) Mandibles

(*c*) Maxillae (*d*) Labium

70. The first pair of jaws, laterally moving, anterior to the maxillae are called

(*a*) Labrum (*b*) Mandibles

(*c*) Maxillae (*d*) Labium

71. The secondary jaws, laterally moving mouth parts behind the mandibles are known as

(*a*) Labrum (*b*) Mandibles

(*c*) Maxillae (*d*) Labium

72. A structure arising from the front surface of the labium in insect mouthparts and which act as a tongue is also known as

(*a*) Hypopharynx (*b*) Labrum

(*c*) Mandibles (*d*) Maxillae

73. Maxilla are complicated and consists of which of the following sclerite

(*a*) Cardo (*b*) Stipes

(*c*) Palpifer (*d*) All of these

74. The first sclerite joins the maxilla to head is

(*a*) Cardo (*b*) Stipes

(*c*) Palpifer (*d*) None of these

75. The second sclerite of maxilla articulates with cardo is known as

(*a*) Scape (*b*) Stipes

(*c*) Palpifer (*d*) None of these

76. On the distal end of the stipes, there are two lobes, the outer lobe is called

(*a*) Galea (*b*) Lacinia

(*c*) Palpifer (*d*) None of these

77. On the distal end of the stipes, there are two lobes, the toothed inner lobe is called

(*a*) Galea (*b*) Lacinia

(*c*) Palpifer (*d*) None of these

78. Salivary gland duct opens through

(*a*) Hypopharynx (*b*) Labrum

(*c*) Mandibles (*d*) Labium

79. The large basal sclerite of the labium is called as

(*a*) Sub mentum (*b*) Mentum

(*c*) Prementum (*d*) None of these

80. The middle sclerite between sub-mentum and prementum is known as

(*a*) Sub mentum (*b*) Mentum

(*c*) Prementum (*d*) None of these

81. The Lepidopteran family which has only 4 functional legs

(*a*) Arctiidae (*b*) Nymphalidae

(*c*) Noctuidae (*d*) Gelichiidae

82. Which one of the following is an auditory membrane?

(*a*) Spiracles (*b*) Pedicel

(*c*) Cranium (*d*) Tympanum

83. The apical sclerite of labium is known as

(*a*) Sub mentum (*b*) Mentum

(*c*) Prementum (*d*) None of these

84. Distally prementum bears two pairs of lobes, the outer pair of lobe is known as

(*a*) Para glossae (*b*) Glossae

(*c*) Ligula (*d*) None of these

85. Distally prementum bears two pairs of lobes, the inner pair of lobe is known as

(*a*) Para glossae (*b*) Glossae

(*c*) Ligula (*d*) None of these

86. Generally prementum bears two pairs of lobes, when both pairs of lobe of prementum fused together is known as

(*a*) Para glossae (*b*) Glossae

(*c*) Ligula (*d*) None of these

87. In piercing and sucking type of mouth parts both mandibles and maxillae are modified into long slender sclerotized hair like structure is called

(*a*) Stylets (*b*) Flabellum

(*c*) Groove (*d*) None of these

88. Within the groove of piercing and sucking type of mouth parts four stylets (fascicle) are protected by

(*a*) Labium (*b*) Labium

(*c*) Mandible (*d*) Maxillae

89. In female mosquito's mouth parts how many stylets are found

(*a*) Four (*b*) Five

(*c*) Six (*d*) Three

90. The proboscis of sponging type of mouthparts consists of

(*a*) Labrum (*b*) Mandible

(*c*) Maxillae (*d*) Labium

91. In sponging type of mouthparts the end of the proboscis is enlarged, sponge like, act as suction pads is called

(*a*) Labella (*b*) Mandible

(*c*) Maxillae (*d*) Labium

92. Siphoning type of mouth parts consist of proboscis which is formed by

(*a*) Two galeae of maxillae (*b*) Tooth of mandible

(*c*) Glossa of labium (*d*) None of these

93. A line of weakness (ecdysial line) on the head above the antennae, inverted "Y" shaped suture found medially on the top of head is called________ suture.

(*a*) Epistomal (*b*) Fronto-clypeal
(*c*) Epicranial or Frontal (*d*) None of these

94. Epistomal suture or Fronto-clypeal suture found between

(*a*) Frons and Vertex (*b*) Frons and Clypeus
(*c*) Vertex and Gena (*d*) None of these

95. Clypeo-labral suture found between

(*a*) Frons and Vertex (*b*) Frons and Clypeus
(*c*) Vertex and Gena (*d*) Clypeus and Labrum

96. The head of insects attached or articulated to the thorax through

(*a*) Pronotum (*b*) Cervix
(*c*) Post-occipital suture (*d*) None of these

97. The thorax of an insect are segmented and consist of

(*a*) Prothorax (*b*) Mesothorax
(*c*) Metathorax (*d*) All of these

98. Mesothorax and Metathorax which bear wing are called

(*a*) Pterostigma (*b*) Pterothorax
(*c*) Pteromere (*d*) None of these

99. Dorsal body plate of thoracic segments are called

(*a*) Tergum or Nota (*b*) Sternum
(*c*) Pleuron (*d*) None of these

100. Ventral body plate of thoracic segments are called

(*a*) Tergum or Nota (*b*) Sternum
(*c*) Pleuron (*d*) None of these

101. Lateral body plate of thoracic segments are called

(*a*) Tergum or Nota (*b*) Sternum
(*c*) Pleuron (*d*) None of these

102. The main functions of thorax concerned with

(*a*) Digestion (*b*) Vision
(*c*) Locomotion (*d*) Growth

103. An insect usually bears three pairs of legs which are located on each of_________thoracic segments?

(*a*) Notum (*b*) Sternum

(*c*) Pleuron (*d*) None of these

104. The first segment of the leg of an insect is articulates with the thorax, called as

(*a*) Coxa (*b*) Trochanter

(*c*) Femur (*d*) Tibia

105. The second segment of the leg of an insect is

(*a*) Trochanter (*b*) Coxa

(*c*) Tarsus (*d*) Pre-tarsus

106. The powerful segment of insect leg which is closely attached to the trochanter is known as

(*a*) Femur (*b*) Coxa

(*c*) Tarsus (*d*) Pre-tarsus

107. The fourth segment of an insect's leg between the femur and the tarsus is

(*a*) Tibia (*b*) Coxa

(*c*) Tarsus (*d*) Pre-tarsus

108. The second segment of the antenna

(*a*) Trochanter (*b*) Pedicel

(*c*) Coxa (*d*) Tarsus

109. In collembolans, jumping organ (furca) is located on the_________abdominal segment.

(*a*) 1st (*b*) 2nd

(*c*) 3rd (*d*) 4th

110. The fifth divisional segment of an insect's leg is

(*a*) Tibia (*b*) Coxa

(*c*) Tarsus (*d*) Pre-tarsus

111. Tarsus of an insect leg is generally subdivided into a number of segments known as

(*a*) Tarsomeres (*b*) Tibia

(*c*) Coxa (*d*) Pre-tarsus

112. Abdominal legs of caterpillar also known as

(*a*) Prolegs (*b*) True leg

(*c*) Artificial legs (*d*) None of these

113. Claspers are the pair of proleg present on__________abdominal segment.

(a) First (b) Second

(c) Forth (d) Last

114. The tip of prolegs is called planta on which hooks or claws are present, which are known as

(a) Claspers (b) Crochets

(c) Crawler (d) Clinger

115. Wings of insects are present in

(a) Larval stage (b) Adult stage

(c) Pupal stage (d) Both (a) & (b)

116. Ectoparasites like head louse, poultry louse and flea insects are

(a) Primarily wingless (b) Secondarily wingless

(c) Deciduous winged (d) None of these

117. __________pairs of wings are visible in true flies.

(a) One (b) Two

(c) Three (d) None of these

118. In hemimetabolus insects (incomplete metamorphosis) wings developed

(a) Externally (b) Internally

(c) Partially (d) None of these

119. The leathery fore wings of an insect known as

(a) Tegmina (b) Elytra

(c) Hemelytra (d) Haltere

120. Thickened horney fore wings of insect, covering and protecting the membranous hind wings known as

(a) Tegmina (b) Elytra

(c) Hemelytra (d) Haltere

121. The basal half of the wing is thick and leathery and distal half is membranous, not involved in flight and protective in nature is known as

(a) Tegmina (b) Elytra

(c) Hemelytra (d) Haltere

122. In true flies the hind wings are modified into small knobbed vibrating organs called

(a) Tegmina (b) Elytra

(c) Hemelytra (d) Haltere

123. The reduced forewings of strepsiptera known as

(a) Pseudohalteres (b) Elytra
(c) Hemelytra (d) Haltere

124. Wings are usually reduced in size, wing margins are fringed are found in

(a) Aphid (b) Thrips
(c) White fly (d) Jassid

125. Wings are covered with small coloured scales founds in

(a) Aphid (b) Thrips
(c) White fly (d) Moths and butterflies

126. The anterior margin strengthened by the costa, called

(a) Coastal margin (b) Apical margin
(c) Anal margin (d) None of these

127. The lateral margin of insect wing is called

(a) Coastal margin (b) Apical margin
(c) Anal margin (d) None of these

128. The posterior margin of insect wing is called

(a) Coastal margin (b) Apical margin
(c) Anal margin (d) None of these

129. The angle by which the wing is attached to the thorax is called

(a) Humeral angle (b) Apical angle or Apex
(c) Anal angle (d) None of these

130. The angle between the coastal and apical margin of insect wing is called

(a) Humeral angle (b) Apical angle or Apex
(c) Anal angle (d) None of these

131. The angle between the apical and anal margin of insect wing is called

(a) Humeral angle (b) Apical angle or Apex
(c) Anal angle (d) None of these

132. An opaque spot near the coastal margin of the wing is called

(a) Pterothorax (b) Pterostigma
(c) Elytra (d) None of these

133. The principal longitudinal veins arranged in order from the anterior margin are

(a) Costa (C) - Radius (R) - Median (M) - Sub-costa (Sc) - Cubitus (Cu) and Anal (A) veins

(b) Costa (C) - Sub-costa (Sc) - Median (M) - Radius (R) - Cubitus (Cu) and Anal (A) veins

(c) Costa (C) - Radius (R) - Sub-costa (Sc) - Median (M) - Cubitus (Cu) and Anal (A) veins

(d) Costa (C) - Sub-costa (Sc) - Radius (R) - Median (M) - Cubitus (Cu) and Anal (A)

134. In honey bee a row of small hooks is present on the costal margin of the hind wing is known as

(a) Bristle (b) Hamuli

(c) Spine (d) Hairs

135. Coupling of wing is achieved by broad overlapping of adjacent margins is known as__________wing coupling.

(a) Amplexiform type (b) Jugate type

(c) Frenate type (d) Hamulate type

136. In strepsiptera Jugam of the fore wings are lobe like and it is locked to the coastal margin of the hind wings is known as

(a) Amplexiform type (b) Jugate type

(c) Frenate type (d) Hamulate type

137. Hind wings of hawk moths, near the base of the coastal margin bears a stout bristle called

(a) Frenulum (b) Bristle

(c) Hamuli (d) Spine

138. Primitive (less advanced) type of insect head

(a) Hypognathous (b) Prognathous

(c) Ophisthognathous (d) None of these

139. The largest part of insect leg is

(a) Trochanter (b) Femur

(c) Tibia (d) Tarsus

140. Frenulum held the retinaculum of forewing arises from sub-coastal veins in case of

(a) Amplexiform type (b) Jugate type

(c) Frenate type (d) Hamulate type

141. Amplexiform type of wing coupling found in

(*a*) Papilionidae and bombycidae (*b*) Apidae and bombidae

(*c*) Aphididae and Jassidae (*d*) None of these

142. The abdominal segments of insects are called

(*a*) Notum (*b*) Uromeres

(*c*) Coxal (*d*) None of these

143. Function of insect head is

(*a*) Food ingestion (*b*) Sensory perception

(*c*) Coordination of bodily activities (*d*) All of these

144. Cranial area of insect head below the frons to which sclerite labrum is attached is known as

(*a*) Vertex (*b*) Occiput

(*c*) Clypeus (*d*) Gena

145. How many abdominal segments are present in spring tail

(*a*) Four (*b*) Six

(*c*) Five (*d*) Eight

146. 2 to 5 abdominal segments are visible and segments 6 to 9 are telescoped within other found in

(*a*) Honey bee (*b*) House fly

(*c*) Hawk moth (*d*) Hover fly

147. In ants, bees and wasps the first abdominal segment is fused with metathorax is called as

(*a*) Propodeum (*b*) Petiole

(*c*) Gaster (*d*) None of these

148. The second abdominal segment of ants, bees and wasps forms a narrow__________.

(*a*) Propodeum (*b*) Petiole

(*c*) Gaster (*d*) None of these

149. The posterior part of the abdomen of hymenoptera have a waist like structure called

(*a*) Propodeum (*b*) Petiole

(*c*) Gaster (*d*) None of these

150. Anal papillae of mosquito surrounds the anus of larvae, concerned with

(*a*) Pupal attachment (*b*) Salt regulation

(*c*) Water absorption (*d*) All of these

151. Cronicles or siphunculi found in aphids on dorsum of_________abdominal segments.

(a) First and second (b) Third and fourth

(c) Fifth or sixth (d) Seventh and eighth

152. A pair of filamentous appendages at the posterior tip of an insect called

(a) Anal styli (b) Caudal style

(c) Cerci or Caudal filament (d) All of these

153. A pair of cerci like unsegmented structure found in last abdominal segment of scolid wasp, are known as

(a) Remigium (b) Pygostyles

(c) Urogomphy (d) Ovipositor

154. A pair of short unsegmented structure found at the end of the abdomen of female cockroach called

(a) Anal cerci (b) Anal styli

(c) Caudal cerci (d) Caudal styli

155. The egg laying organ found in female insect called

(a) Gonopore (b) Urogomphy

(c) Ovipositor (d) Ootheca

156. Ovipositor modified into sting in

(a) Drone bee (b) House fly

(c) Fruit fly (d) Worker honey bee

157. Pseudo ovipositor found in

(a) Fruit flies (b) House fly

(c) Wasp (d) Both (a) & (b)

158. The external sexual organ of male insects called

(a) Gonopore (b) Male genitalia

(c) Female genitalia (d) None of these

159. Male genitalia of insects confined on which segment of the abdomen

(a) 8th (b) 9th

(c) 6th (d) 7th

160. The type of mouth parts found in larval Lepidoptera

(a) Sponging (b) Siphoning

(c) Piercing and Sucking (d) Biting and Chewing

161. In Coleoptera the type of forewing is of extensively sclerotized is called

(*a*) Halteres (*b*) Elytra

(*c*) Hemelytra (*d*) Tegmina

162. Cellular layer of integument is

(*a*) Epicuticle (*b*) Endocuticle

(*c*) Basement membrane (*d*) Epidermis

163. Outermost layer of the insect integument is

(*a*) Cement layer (*b*) Cuticulin layer

(*c*) Both (*a*) & (*b*) (*d*) None of these

164. Which structure is found in an insect pretarsus?

(*a*) Trochanter (*b*) Furca

(*c*) Empodium (*d*) All of these

165. Trochanter is two segmented in

(*a*) Lepidoptera (*b*) Hemiptera

(*c*) Hymenoptera (*d*) Coleoptera

166. Coxal articulation with the thorax is

(*a*) Monocondylic (*b*) Dicondylic

(*c*) Both (*a*) & (*b*) (*d*) None of these

167. Basitarsus is connected with the

(*a*) Pretarsus (*b*) Femur

(*c*) Tibia (*d*) Coxa

168. An ommatidium is best defined as a

(*a*) Sub division of ventral nerve cord

(*b*) Functional unit of the compound eye

(*c*) Ventral lobe of the insect's brain

(*d*) Joint of coxa and trochanter

169. Sucking mouthparts are not found in

(*a*) Fleas (*b*) Flies

(*c*) Lice (*d*) Ants

170. Antenna is articulated with cranium in a single point, the

(*a*) Antennifera (*b*) Frons

(*c*) Gena (*d*) Johnston's organ

171. Mandibles are toothed in which of the following type of mouthparts

(a) Biting and chewing (b) Chewing and lapping
(c) Rasping and sucking (d) All of these

172. Prologs are distally armed with

(a) Crochets (b) Creeping welt
(c) Spine (d) Seta

173. Glossa and paraglossa are the lobes of

(a) Labium (b) Labrum
(c) Maxillae (d) Mandible

174. Legless and delicate larvae are

(a) Protopod (b) Apodus
(c) Oligopod (d) Polypod

175. In biting and chewing type of mouth parts the upper lip is

(a) Labium (b) Labrum
(c) Hypopharynx (d) Labrum-epipharynx

176. Mouth parts show asymmetry in

(a) Biting and chewing type (b) Sponging type
(c) Siphoning type (d) Rasping and sucking type

177. Major body segments of insects is

(a) Thorax (b) Head
(c) Tagmen (d) All of these

178. Which of the following is not a convex vein?

(a) Costa (b) Radius
(c) Anterior cubitus (d) Posterior cubitus

179. The larva which has no legs and is very poorly sclerotized is called

(a) Oligopod (b) Apodous
(c) Scarabaeiform (d) Polypod

180. Cement layer of insect integument is also known as

(a) Epicuticle (b) Tectocuticle
(c) Endocuticle (d) Both (a) & (b)

181. Prtocerebrum innervates

(a) Compound eye (b) Antennae
(c) Labrum (d) Labium

182. Tympanal organs are located in between metathoracic legs of

(*a*) Mantids (*b*) Dragon fly

(*c*) Grasshopper (*d*) Termite

183. Facets of compound eye are circular in

(*a*) Dragon fly (*b*) Damsel fly

(*c*) Honey bee (*d*) Thrips

184. Visual organs of holometabolous larvae

(*a*) Dorsal ocelli (*b*) Lateral ocelli

(*c*) Stemmata (*d*) Both (*b*) & (*c*)

185. Tenant hairs are seen in

(*a*) Honey bee (*b*) House fly

(*c*) Mosquito (*d*) Moth

186. Number of basic segments of insect leg is

(*a*) 5 (*b*) 6

(*c*) 7 (*d*) 8

187. Femur is the__________segment of insect leg.

(*a*) 1st (*b*) 2nd

(*c*) 3rd (*d*) 4th

188. Coxa is articulated with the

(*a*) Sternum (*b*) Pleuron

(*c*) Tergum (*d*) None

189. For legs of mole crickets are modified into

(*a*) Saltatorial (*b*) Fossorial

(*c*) Natatorial (*d*) Cursorial

190. Orientation of head of aphids with respect to body is

(*a*) Prognathous (*b*) Opisthognathous

(*c*) Hypognathous (*d*) None of these

191. The antenna of termite is

(*a*) Moniliform (*b*) Filliform

(*c*) Clavate (*d*) Clubbed

192. The part of labium from which the labial palp arise is known as

(*a*) Sub-mentum (*b*) Pre-mentum

(*c*) Post-mentum (*d*) None of these

193. Cuticle of insect is

(a) Unicellular
(b) Multicellular
(c) Noncellular
(d) None of these

194. The layer of integument in between procuticle and endocuticle is

(a) Subcuticle
(b) Mesocuticle
(c) Epicuticle
(d) None of these

195. The most common sulcus of insect head is

(a) Frontoclypeal
(b) Circum antennal
(c) Subocular
(d) None of these

196. Exo and endo cuticle are collectively known as

(a) Procuticle
(b) Subcuticle
(c) Epicuticle
(d) None of these

197. Units of compound eye is

(a) Simple eye
(b) Ommatidium
(c) Stemmata
(d) Ocelli

198. The primitive type of mouthparts found in insect is

(a) Chewing and biting
(b) Piercing and sucking
(c) Sponging
(d) Siphoning

199. The largest group of Arthropoda is

(a) Crustacea
(b) Insecta
(c) Arachnida
(d) Onychophora

200. Seta is arises from

(a) Epicuticle
(b) Pore cannal
(c) Alveolus
(d) Basement membrane

201. Flagellomeres are the subsegments of

(a) Pedicel
(b) Scape
(c) Flagellum
(d) None of these

202. Solid cuticular invaginations are known as

(a) Apophysis
(b) Apodeme
(c) Tentorium
(d) Spur

203. Tentorium is an internal cuticular skeleton found on

(a) Head
(b) Thorax
(c) Abdomen
(d) Wing

204. Siphon of siphoning type of mouthparts is

(*a*) Lacinia (*b*) Galea

(*c*) Cardo (*d*) Palpifer

205. The insect cuticle chemically consists of

(*a*) Chitin and fat (*b*) Protein and fat

(*c*) Chitin and protein (*d*) Chitin and glycogen

206. The seta forming cell Trichogen is__________origin.

(*a*) Epidermal (*b*) Endodermal

(*c*) Endodermal (*d*) Both (*a*) & (*b*)

207. Rigid outgrowths of integument are known as

(*a*) Spines (*b*) Cuticular process

(*c*) Cuticular appendage (*d*) None of these

208. Gena is situated on the insect head

(*a*) Laterally (*b*) Dorsally

(*c*) Frontally (*d*) Ventrally

209. The order lack antennae entirely is

(*a*) Protura (*b*) Collembola

(*c*) Neuroptera (*d*) Odonata

210. The glossa and paraglossa are collectively known as

(*a*) Ligula (*b*) Prementum

(*c*) Submentum (*d*) Palpifer

211. In case of piercing and sucking mouthparts the rostrum is dorsally grooved to accommodate

(*a*) Labrum (*b*) Mandible

(*c*) Maxillae (*d*) Both (*b*) & (*c*)

212. Legs are arises from the thoracic segments

(*a*) Laterally (*b*) Ventro-laterally

(*c*) Ventrally (*d*) Dorsally

213. Maximum number of tarsal segments are

(*a*) 5 (*b*) 6

(*c*) 7 (*d*) 4

214. Which of the following is not a head sclerite?

(*a*) Gena (*b*) Vertex

(*c*) Epimeron (*d*) Frons

215. Axillary sclerites are present in

(a) Coleoptera
(b) Lepidoptera
(c) Dictyoptera
(d) Odonata

216. Piecing and sucking type of mouthparts are found in which of the following insects groups

(a) Hemipteroids
(b) Orthopteroids
(c) Holometabolous
(d) All of these

217. What type of chemical monomer forms the backbone of a chitin molecule?

(a) Lipid
(b) Quinone
(c) Sugar
(d) Amino Acid

218. An insect's mechanoreceptors would not be sensitive to

(a) Body movement
(b) Wind speed
(c) Sound vibrations
(d) Water vapor

219. Stridulation is a method of producing sound by

(a) Vibrating the wings
(b) Vibrating a resonant membrane
(c) Striking the substrate
(d) Rubbing body parts together

220. In Dictyoptera and Orthoptera, the forewing which appears to be extensively sclerotized is called

(a) Halteres
(b) Elytra
(c) Hemielytra
(d) Tegmina

221. The epistomal suture is absent in

(a) *Blatta orientalis*
(b) Grasshopper
(c) Gryllids
(d) Mantids

222. To which body segment are the halters attached

(a) Mesothorax
(b) First abdominal segment
(c) Prothorax
(d) Metathorax

223. What is the maximum number of ocelli that may be found in an adult insect

(a) Zero
(b) Three
(c) Five
(d) Ten

224. Which mouthparts lies between the labrum and the maxillae

(a) Hypopharynx
(b) Mandibles
(c) Labium
(d) Palps

225. Which sclerite lies below (ventral to) the frons?

(*a*) Gena (*b*) Clypeus

(*c*) Occiput (*d*) Labium

226. A tormogen cell is always associated with a

(*a*) Spine (*b*) Gland

(*c*) Apodeme (*d*) Seta

227. Which part of the exoskeleton lies between the wax layer and the cement layer?

(*a*) Exocuticle (*b*) Endocuticle

(*c*) Cuticulin layer (*d*) None of these

228. Elastic regions of the exoskeleton are

(*a*) Generally known as sclerites

(*b*) Found only at the joints

(*c*) Lack a well defined exocuticle

(*d*) Contain high concentrations of quinones

229. Which layer (s) of the exoskeleton is (are) secreted by the epidermis?

(*a*) Exocuticle (*b*) Endocuticle

(*c*) Epicuticle (*d*) All of these

230. The pedicel is the name for the

(*a*) 1st leg segment (*b*) 2nd leg segment

(*c*) 1st antennal segment (*d*) 2nd antennal segment

231. All stinging insects belong to the order

(*a*) Hemiptera (*b*) Thysanoptera

(*c*) Hymenoptera (*d*) Diptera

232. In insects, the first pair of post-oral appendages is called mandibles. What are these appendages called in Arachnids

(*a*) Maxillae (*b*) Antennae

(*c*) Walking legs (*d*) Chelicerae

233. Which sclerite lies below the epistomal suture?

(*a*) Frons (*b*) Labrum

(*c*) Clypeus (*d*) Gena

234. In piercing and sucking type of mouth parts, food channel is formed by

(*a*) Maxillary palpi (*b*) Mandibles

(*c*) Stylets (*d*) Galeae

235. Saltatorial Legs are found in

(a) Cockroach (b) Praying mantid

(c) Grasshopper (d) Mole cricket

236. Chewing mouthparts are not found in

(a) Crickets (b) Weevils

(c) Caterpillar (d) Stoneflies

237. Insects without compound eyes are belongs to which of the following groups

(a) Diptera and Hemiptera (b) Protura and Diplura

(c) Lepidoptera and Coleoptera (d) None of these

238. The fore wing of Strepsiptera is called

(a) Pseudo haltare (b) Tegmina

(c) Elytra (d) Hemelytra

239. Cranial area between occipital and post occipital suture is__________.

(a) Occiput (b) Gena

(c) Fons (d) Clypeus

240. Bipectinate type of antennae found in

(a) Grasshopper (b) Honey bee

(c) Moths (d) Male mango mealy bug

241. Hypognathus type of head is found in

(a) Bugs (b) Beetles

(c) Weevils (d) Grasshoppers

242. Trochanter is a part of

(a) Antennae (b) Wings

(c) Mouth (d) Leg

243. Lacinia is a part of

(a) Labrum (b) Mandible

(c) Maxilla (d) Labium

244. Scansorial type of leg is found in

(a) Louse (b) Cricket

(c) Ant (d) Grasshopper

245. Which part of maxillae is modified to form a coiled spring in siphoning type of mouth parts?

(a) Cardo (b) Stipes

(c) Galea (d) Lacinia

246. Brachypterous type of wings are found in

(*a*) Grasshopper (*b*) Housefly

(*c*) Butterflies (*d*) Aphid

247. In Grasshopper the tympanal organs is situated in_________abdominal tergum.

(*a*) First (*b*) Second

(*c*) Third (*d*) Sixth

248. Larvae of housefly is called

(*a*) Maggot (*b*) Grub

(*c*) Nymph (*d*) Caterpillar

249. The larvae of mustard sawfly are

(*a*) Apodus (*b*) Curculionoid

(*c*) Polypod (*d*) None of these

250. Decticous pupa are found in which of the following order

(*a*) Lepidoptera (*b*) Coleoptera

(*c*) Neuroptera (*d*) Hymenoptera

251. Exarate adecticous pupae are mostly found in order

(*a*) Lepidoptera (*b*) Coleoptera

(*c*) Diptera (*d*) Hymenoptera

252. Scarabaid form larvae are found in order

(*a*) Lepidoptera (*b*) Coleoptera

(*c*) Diptera (*d*) Hymenoptera

253. Opisthognathous type of mouth parts are found in

(*a*) Bugs (*b*) Beetles

(*c*) Aphids (*d*) Grasshoppers

254. Prognathous type of mouth parts are found in

(*a*) Bugs (*b*) Beetles

(*c*) Aphids (*d*) Grasshoppers

255. Tentorium is located in the

(*a*) Head (*b*) Thorax

(*c*) Abdomen (*d*) All of these

256. The example of insect having sharp ovipositor

(*a*) Beetle (*b*) Jassids

(*c*) Mustard saw fly (*d*) Aphids

257. The meaning of term 'Propodium' is

(a) Remaining part of the abdomen
(b) Waist of the insects
(c) Part of the alimentary canal
(d) Fused segments of the abdomen

258. The type of mouth parts found in adult butterflies and moths

(a) Sponging (b) Siphoning
(c) Piercing and Sucking (d) Biting and Chewing

259. In siphoning type of mouth parts watch spring type (proboscis) of structure is modification of

(a) Maxillary palpi (b) Mandibles
(c) Stylets (d) Galeae

260. Hind wing of housefly is modified into

(a) Air wings (b) Elytra
(c) Hymen (d) Halters

261. Function of epicuticle is

(a) Prevent loss of water from insects integument
(b) Prevent loss of nutrient from leaf surface
(c) Prevent loss of air from surface
(d) Prevent wax secretion

262. Aristate type of antennae is found in

(a) Mosquito (b) Housefly
(c) Cockroach (d) Honeybee

263. Setaceous type of antennae is found in

(a) Mosquito (b) Housefly
(c) Cockroach (d) Honeybee

264. Filiform type of antennae is found in

(a) Grasshopper (b) Bugs
(c) Cricket (d) Honeybee

265. Serrate type of antennae is found in

(a) Pulse beetle (b) Coconut beetle
(c) Grasshopper (d) Honeybee

266. Pectinate type of antennae is found in

(*a*) Gall fly
(*b*) Rice hopper
(*c*) Root borer
(*d*) Saw fly

267. Bipectinate type of antennae is found in

(*a*) Butterfly
(*b*) Aphid
(*c*) Silkmoth
(*d*) Honeybee

268. Flabellate type of antennae is found in

(*a*) Aphid
(*b*) Cedar beetle
(*c*) Grasshopper
(*d*) Honeybee

269. Plumose type of antennae is found in

(*a*) Female Mosquito
(*b*) Male Housefly
(*c*) Aphids
(*d*) Male Mosquito

270. Whorled type of antennae is found in

(*a*) Termite
(*b*) Male mango mealy bug
(*c*) Aphid
(*d*) Grasshopper

271. Clavate type of antennae is found in

(*a*) Moths
(*b*) Butterfly
(*c*) Aphid
(*d*) Honeybee

272. Capitate type of antennae is found in

(*a*) Coleoptera and Lepidoptera
(*b*) Homoptera and Heteroptera
(*c*) Orthoptera and Thysanoptera
(*d*) Lepidoptera and Hymenoptera

273. Lamellate type of antennae is found in

(*a*) Ground beetle
(*b*) Flea beetle
(*c*) Scarabaeid beetle
(*d*) None of these

274. Stylate type of antennae is found in

(*a*) Mosquito
(*b*) Housefly
(*c*) Horse fly
(*d*) None of these

275. Fusiform type of antennae is found in

(*a*) Mosquito
(*b*) Housefly
(*c*) Silkmoth
(*d*) Sphingid moth

276. Propodeum is a segments of

(a) Head (b) Thorax

(c) Abdomen (d) Wing

277. __________type of mouth parts occurs in housefly.

(a) Chewing and Lapping (b) Siphoning

(c) Sponging (d) Chewing and Biting

278. __________type of mouth parts occurs in honeybees.

(a) Chewing and Lapping (b) Siphoning

(c) Sponging (d) Chewing and Biting

279. Cursorial type of legs are found in

(a) Grasshopper (b) Ants

(c) Cockroach (d) House fly

280. Scansorial type of legs are found in

(a) Head louse (b) Wasps

(c) Moth (d) House fly

281. Saltatorial type of legs are found in

(a) Bugs (b) Beetles

(c) Grasshopper (d) House fly

282. Fossorial or digging type of legs are found in

(a) Mole cricket (b) Wasps

(c) Aphids (d) House fly

283. Natatorial or swimming type of legs are found in

(a) Giant water bugs (b) Wasps

(c) Cockroach (d) House fly

284. Foragial or pollen collecting type of legs are found in

(a) Worker honey bee (b) Queen bee

(c) Drone (d) None of these

285. Raptorial or grasping type of legs are found in

(a) Thrips (b) Praying Mantis

(c) Cockroach (d) House fly

286. Climbing or sticking type of legs are found in

(a) Beetle (b) Wasps

(c) Cockroach (d) House fly

287. Which type of wing modification are found in Orthoptera?

(*a*) Elytra (*b*) Hemelytra

(*c*) Tegmina (*d*) Halters

288. Which type of wing modification are found in Diptera?

(*a*) Elytra (*b*) Hemelytra

(*c*) Tegmina (*d*) Halters

289. Which type of wing modification are found in Strepsiptera?

(*a*) Elytra (*b*) Pseudohalters

(*c*) Brachypterous (*d*) Halters

290. Piercing and Sucking mouthparts are found in

(*a*) Mosquitoes (*b*) Honey bees

(*c*) Thrips (*d*) Butterflies

291. In thrips, the rasping and sucking mouth parts are asymmetrical due to absence of

(*a*) Right Maxillae (*b*) Right Mandible

(*c*) Left Mandible (*d*) Labrum

292. Which type of wing modification are found in Hemipteran insects?

(*a*) Elytra (*b*) Hemelytra

(*c*) Brachypterous (*d*) Halters

293. Hypothetical type of wing venation was proposed by

(*a*) Srivastava and Pandey (*b*) Pruthi and Puri

(*c*) Comstock and Needham (*d*) Imms

294. Medius vein is absent in which order

(*a*) Hemiptera (*b*) Hymenoptera

(*c*) Diptera (*d*) Thysanoptera

295. Humeral wing angle is located at

(*a*) Base of costal vein (*b*) End of radius vein

(*c*) End of Medius vein (*d*) End of subcostal vein

296. Jugal and humeral lobe type of wing coupling apparatus is found in

(*a*) Coleoptera (*b*) Hymenoptera

(*c*) Diptera (*d*) Lepidoptera

297. Frenulum and retinaculum type of wing coupling apparatus occur in

(*a*) Lepidoptera (*b*) Orthoptera

(*c*) Hemiptera (*d*) Coleoptera

298. Amplexiform type of wing coupling apparatus occur in

(*a*) Lepidoptera (*b*) Hymenoptera

(*c*) Hemiptera (*d*) Coleoptera

299. Jugal wing coupling apparatus occur in

(*a*) Lepidoptera (*b*) Thysanoptera

(*c*) Trichoptera (*d*) Coleoptera

300. Hook and channel type of wing coupling apparatus occur in

(*a*) Lepidoptera (*b*) Orthoptera

(*c*) Hemiptera (*d*) Coleoptera

301. Female reproductive organs are found on__________abdominal segment.

(*a*) 8th (*b*) 9th

(*c*) Between 8th and 9th (*d*) None of these

302. Male reproductive organs are found on__________abdominal segment.

(*a*) 8th (*b*) 9th

(*c*) Between 8th and 9th (*d*) None of these

303. The larvae which have no locomotory organs generally found in which of the following group

(*a*) Nymph (*b*) Maggot

(*c*) Caterpillar (*d*) Grub

304. Which of the following muscles are directly attached to wing?

(*a*) Dorsal muscles (*b*) Axillary muscles

(*c*) Basalar muscles (*d*) Sub-alar muscles

305. Example of lateral sclerite is

(*a*) Basisternum (*b*) Episternum

(*c*) Sternum (*d*) Pre-sternium

306. Which of the following is an abdominal appendage?

(*a*) Cerci (*b*) Waist

(*c*) Gonopore (*d*) Apophysis

307. The terminal abdominal segments 6 to 9 are telescoped within the segment 5 when not in use. This condition is found in

(*a*) Gall fly (*b*) House fly

(*c*) Dragon fly (*d*) White fly

308. The composition of insect integument is

(*a*) Cuticle + Basement membrane

(*b*) Cuticle + Epidermis + Basement membrane

(*c*) Cuticle + Epidermis

(*d*) Epidermis + Basement membrane

309. Example of 1st longitudinal vein is

(*a*) Costal vein (*b*) Radial vein

(*c*) Median vein (*d*) None of these

310. Pronotum is easily seen in which insects

(*a*) Ants (*b*) Termites

(*c*) Cockroach (*d*) Plant bugs

311. Stylets are parts of mouthparts found in

(*a*) Ants (*b*) Termites

(*c*) Cockroach (*d*) Plant bugs

312. Sting is modified part of

(*a*) Mouth part (*b*) Thorax

(*c*) Ovipositor (*d*) Wings

313. On the last segment of the abdomen of male cockroach having a pair of

(*a*) Anal styli (*b*) Caudal styli

(*c*) Claspers (*d*) Caudal filament

314. Head and metathorax fused with each other is called

(*a*) Cephalothorax (*b*) Mandible

(*c*) Antennae (*d*) Mesopodeum

315. A promising character of the order Odonata is Pterostigma which is found on

(*a*) Wing (*b*) Legs

(*c*) Antennae (*d*) Eye

316. Degenerate ocelli of cockroach is known as

(*a*) Cylassa (*b*) Acrotergite

(*c*) Fenestra (*d*) None of these

317. The males are winged and females are wingless in

(*a*) Coccids (*b*) Embioptera

(*c*) Strepsiptera (*d*) All of these

318. In which of the following insect order, both pairs of wings are lost completely providing apterous condition?

(a) Psocoptera (b) Mallophaga
(c) Siphonaptera (d) All the above

319. Adult insect possess__________numbers of legs.

(a) 4 (b) 5
(c) 6 (d) 8

320. Adult mite possess__________numbers of legs.

(a) 4 (b) 5
(c) 6 (d) 8

321. Wings of the insects are

(a) Quadriangular (b) Somewhat triangular
(c) Fairly biangular (d) Irregular shape

322. Hexapod creature are

(a) Centipede (b) Mite
(c) Insect (d) Millipede

323. Octapod creature are

(a) Centipede (b) Spider
(c) Insect (d) Millipede

324. Exoskeleton of insects consists of

(a) Calcium (b) Protein
(c) Chitin (d) Chitin + Protein

325. Wings are situated in the thorax

(a) Ventro laterally (b) Dorso laterally
(c) Dorsally (d) Ventrally

326. The antennae are appendages of which neuromere

(a) Tritocerebrum (b) Deutocerebrum
(c) Protocerebrum (d) Labial ganglion

327. Indicate which represents the analogous group

(a) Wing of butterfly and birds
(b) Forelegs of horse and arms of man
(c) Forewings of moth and fin of fishes
(d) Hind wings of butterfly and legs of birds

328. The sclerites present at wing base for jointing thorax are collectively called as

(*a*) Pteralia (*b*) Fastigium

(*c*) Jegulae (*d*) Scutellum

329. Halters which maintain balance during flight are present in

(*a*) Lepidoptera (*b*) Diptera

(*c*) Dermaptera (*d*) Hemiptera

330. Pollen basket is located on

(*a*) Hind leg (*b*) Fore leg

(*c*) Both (*a*) & (*b*) (*d*) None of these

331. Johnston's organ is located in the

(*a*) Legs (*b*) Wings

(*c*) Mouth parts (*d*) Antennae

332. Difference in male and female cockroach is

(*a*) Shape of wing

(*b*) Number of antennal segments

(*c*) Structure of compound eyes

(*d*) Presence or absence of anal style

333. Monocondylic single mandible is found in

(*a*) Hemiptera (*b*) Hymenoptera

(*c*) Diptera (*d*) Thysanoptera

334. Laval prolegs are occur on which abdominal segment

(*a*) 3 to 5 and 7 (*b*) 3 to 5 and 8

(*c*) 3 to 6 and 9 (*d*) 3 to 6 and 10

335. How many pair of legs are found in mustard saw fly at larval stage?

(*a*) 4 pairs (*b*) 8 pairs

(*c*) 6 pairs (*d*) 10 pairs

336. In aphids cornicals are located on__________abdominal segment.

(*a*) 4th (*b*) 5th

(*c*) 6th (*d*) 7th

337. Newly emerged, active immature forms of a scale insect are called

(*a*) Claspers (*b*) Naiads

(*c*) Crawler (*d*) Larva

338. A pair of sensory appendages present at 11th abdominal segment of insect is known as

(*a*) Paraproct (*b*) Caudal styli

(*c*) Cerci (*d*) None of these

339. The pupal stage in most butterflies that do not construct cocoons is called

(*a*) Pharate (*b*) Chrysalis

(*c*) Imago (*d*) Creep

340. A spine like structure found at the tip of the abdomen of a chrysalis is called

(*a*) Caudal styli (*b*) Cerci

(*c*) Cremaster (*d*) None of these

341. A tube like structure found on the ventral side of the first abdominal segment of Collembolans is

(*a*) Gladiator (*b*) Colophore

(*c*) Fontanili (*d*) None of these

342. The outer surface or lens of an ommatidium is called as

(*a*) Lash (*b*) Retina

(*c*) Cornea (*d*) None of these

343. The terminal body segment in some primitive arthropods is

(*a*) Wax gland (*b*) Olfactory tube

(*c*) Telson (*d*) None of these

344. A clasplike structure associated with the furcula of Collembola is called

(*a*) Telson (*b*) Tenaculum

(*c*) Triungulin (*d*) Tentorium

345. An active, first instar of insects that undergo hyper metamorphosis is called

(*a*) Telson (*b*) Tenaculum

(*c*) Triungulin (*d*) Tentorium

346. A pad like structure present at the base of each tarsal claw is known as

(*a*) Crochets (*b*) Pretarsus

(*c*) Pulvillus (*d*) Empodium

347. The differentiation of the body of insects into distinct functional regions is called as

(*a*) Morphogenesis (*b*) Oogenesis

(*c*) Tagmosis (*d*) Blastogenesis

348. Wax layer of insect cuticle is closely packed with wax molecule prevents escape of

(*a*) Polyphenols (*b*) Water

(*c*) Lipid (*d*) Air

349. Nodus wing is present in the insect order

(*a*) Odonata (*b*) Collembola

(*c*) Diptera (*d*) Protura

Fill in the Blanks

1. A simple eye of insects and other arthropods incorporating a single lens is known as__________.

2. Compound eye of insects consist of several ommatidia, each of which functions like an__________.

3. Each eye is surrounded by a narrow ring like__________.

4. Except scape and pedicel, the remaining divisions of antennae together constitute the flagellum, this part is also known as__________.

5. Gnathal appendages and lobes of the head surrounding the oral aperture of the alimentary canal collectively known as__________.

6. A typical mouth parts of insect consists of a labrum (upper lip), a pair of mandibles, a pair of maxillae, a labium (lower lip) and a__________.

7. __________covers the mouth cavity from above and it helps to pull the food in to the mouth.

8. Labium is the lower lip, it bounds the__________from below or behind.

9. Labium consists of three median sclerites, prementum, mentum and__________.

10. In biting and chewing type of mouthparts inner surface of labrum is referred to as__________, it is an organ of taste.

11. The mandible acts transversely to__________the food into small pieces.

12. In rasping and sucking type of mouth parts, mouth cone consists of Labrum,__________and maxillae.

13. Three stylets of rasping and sucking type of mouth parts derived from two maxillae and__________.

14. Stylets of rasping and sucking type of mouth parts are useful to lacerate the plant tissue and the oozing sap is sucked up by the__________.

15. In chewing and lapping type of mouth parts maxillolabial structures are modified to form the__________.

16. In chewing and lapping type of mouth parts the glossa terminates into a small circular spoon shaped lobe called spoon or bouton or flabellum which is useful to__________.

17. The proboscis of sponging type of mouth parts differentiated into basal rostrum and distal__________.

18. Primitive insects like silverfish, springtail have no wings are called__________.

19. In ants and termites wings are__________.

20. Wings are moved by__________attached to their base.

21. The wing of insect is nourished by blood circulating through__________.

22. Presence of venation in wings of insects is an useful parameters for the purpose of__________.

23. The most widely used system of nomenclature for the veins of the wing is__________.

24. Small veins often found inter connecting the longitudinal veins are called__________.

25. A "Y" shaped organ present in the ventral side of the fourth abdominal segment, it exerts a force against the substratum and the insect is propelled in the air, called__________.

26. Aquatic immature stages of hemimetabolous insects are called as__________.

27. In insect abdomen eight pairs of spiracles are present on__________abdominal segments.

28. In grasshopper a pair of auditory organ Tympanum is found one on either side of the__________.

29. The tympanum are connected to the metathoracic ganglia through__________.

30. In queen termite after mating the abdomen become gradually swollen due to the enlargement of ovaries called__________.

31. What is aedeagus__________.

32. Short unjointed abdominal appendages of silverfish, which are reduced abdominal legs helps in locomotion, is known as__________.

33. The wing is triangular in shape and has three sides and three__________.

34. The basal tarsal is often larger than other is named as__________.

Match the Following

1. Match the cross veins joins the longitudinal veins are

(a) Humeral cross vein	i) m_2 and m_3 branches of median
(b) Radio - medial cross vein	ii) Sub-radius and median
(c) Medial cross vein	iii) Medius and cubitus
(d) Medio-cubital cross vein	iv) Costa to Sub-costa

2. Match the types of antennae are found in insects

(a) Female mosquito	i) Plumose
(b) Male mosquito	ii) Pilose
(c) Honey bee	iii) Aristate
(d) House fly	iv) Geniculate

3. Match the types and modifications of legs

(a) Ambulatorial	i) Jumping leg
(b) Cursorial	ii) Walking leg
(c) Saltatorial	iii) Climbing leg
(d) Scansorial	iv) Running leg

4. Match the types and modifications of legs found in insects

(a) Fossorial	i) Collect food materials
(b) Raptorial	ii) Swimming leg
(c) Natatorial	iii) Digging leg
(d) Foragial legs	iv) Grasping leg

5. Match the types of modifications of legs found in insects

(a) Sticking leg	i) Water beetle
(b) Basket like leg	ii) Dragon fly
(c) Clasping legs	iii) House fly
(d) Natatorial	iv) Water bugs

6. **Match the types of mouth parts found in insects**

(a)	Siphoning type	i)	Butterfly and Moths
(b)	Chewing and Lapping type	ii)	Thrips
(c)	Rasping and sucking type	iii)	Honey bee
(d)	Sponging type	iv)	House fly

7. **Match the type of insect antennae with their character**

(a)	Moniliform	i)	Saw like appearance
(b)	Setaceous	ii)	Tapers gradually to the tip
(c)	Serrate	iii)	String of beads
(d)	Pectinate	iv)	Appearance like a comb

8. **Match the type of insect antennae with their character**

(a)	Clavate	i)	Club shaped tip
(b)	Lamellate	ii)	Appearance of a feather
(c)	Geniculate	iii)	Elbow like
(d)	Plumose	iv)	Fan like appearance

9. **Match the type of insect mouthparts found in insects**

(a)	Biting and chewing	i)	Larval Lepidoptera
(b)	Piercing and sucking	ii)	Mosquitoes
(c)	Sponging	iii)	Housefly
(d)	Siphoning	iv)	Adult Lepidoptera

10. **Match the following type of wing found in insect**

(a)	Tegmina	i)	Beetle
(b)	Elytra	ii)	Grasshopper
(c)	Haltere	iii)	Thrips
(d)	Fringed wings	iv)	House fly

Answers

Multiple choice questions

1	(b)	21	(d)	41	(c)	61	(d)
2	(a)	22	(b)	42	(d)	62	(c)
3	(d)	23	(d)	43	(a)	63	(c)
4	(a)	24	(d)	44	(b)	64	(c)
5	(a)	25	(a)	45	(d)	65	(b)
6	(c)	26	(a)	46	(c)	66	(c)
7	(c)	27	(a)	47	(b)	67	(c)
8	(d)	28	(a)	48	(d)	68	(a)
9	(a)	29	(b)	49	(b)	69	(d)
10	(a)	30	(d)	50	(b)	70	(b)
11	(b)	31	(d)	51	(d)	71	(c)
12	(b)	32	(b)	52	(d)	72	(a)
13	(d)	33	(a)	53	(a)	73	(d)
14	(c)	34	(b)	54	(b)	74	(a)
15	(c)	35	(c)	55	(c)	75	(b)
16	(b)	36	(c)	56	(d)	76	(a)
17	(d)	37	(a)	57	(a)	77	(b)
18	(d)	38	(b)	58	(a)	78	(a)
19	(a)	39	(a)	59	(a)	79	(a)
20	(d)	40	(b)	60	(c)	80	(b)

81	(b)	115	(b)	149	(c)	183	(c)
82	(d)	116	(a)	150	(b)	184	(d)
83	(c)	117	(a)	151	(c)	185	(b)
84	(a)	118	(a)	152	(a)	186	(a)
85	(b)	119	(a)	153	(b)	187	(c)
86	(c)	120	(b)	154	(a)	188	(b)
87	(a)	121	(c)	155	(c)	189	(b)
88	(b)	122	(d)	156	(d)	190	(b)
89	(c)	123	(a)	157	(d)	191	(a)
90	(d)	124	(b)	158	(b)	192	(b)
91	(a)	125	(d)	159	(b)	193	(c)
92	(a)	126	(a)	160	(d)	194	(b)
93	(c)	127	(b)	161	(b)	195	(a)
94	(b)	128	(c)	162	(d)	196	(a)
95	(d)	129	(a)	163	(a)	197	(b)
96	(b)	130	(b)	164	(c)	198	(a)
97	(d)	131	(c)	165	(c)	199	(b)
98	(b)	132	(b)	166	(a)	200	(c)
99	(a)	133	(d)	167	(c)	201	(c)
100	(b)	134	(b)	168	(b)	202	(a)
101	(c)	135	(a)	169	(d)	203	(a)
102	(c)	136	(b)	170	(b)	204	(b)
103	(c)	137	(a)	171	(a)	205	(c)
104	(a)	138	(a)	172	(a)	206	(a)
105	(a)	139	(b)	173	(a)	207	(b)
106	(a)	140	(c)	174	(b)	208	(a)
107	(a)	141	(a)	175	(b)	209	(a)
108	(b)	142	(b)	176	(d)	210	(a)
109	(d)	143	(d)	177	(a)	211	(d)
110	(c)	144	(c)	178	(d)	212	(a)
111	(a)	145	(b)	179	(b)	213	(a)
112	(a)	146	(b)	180	(a)	214	(c)
113	(d)	147	(a)	181	(a)	215	(d)
114	(b)	148	(b)	182	(a)	216	(a)

217	(b)	251	(b)	285	(b)	319	(c)
218	(c)	252	(b)	286	(d)	320	(d)
219	(d)	253	(c)	287	(c)	321	(b)
220	(d)	254	(b)	288	(d)	322	(c)
221	(a)	255	(a)	289	(b)	323	(b)
222	(d)	256	(c)	290	(a)	324	(d)
223	(b)	257	(d)	291	(b)	325	(b)
224	(b)	258	(b)	292	(b)	326	(b)
225	(b)	259	(d)	293	(c)	327	(a)
226	(d)	260	(d)	294	(c)	328	(a)
227	(c)	261	(a)	295	(a)	329	(b)
228	(b)	262	(b)	296	(d)	330	(a)
229	(d)	263	(c)	297	(a)	331	(d)
230	(d)	264	(a)	298	(a)	332	(d)
231	(c)	265	(a)	299	(a)	333	(d)
232	(d)	266	(d)	300	(a)	334	(d)
233	(d)	267	(c)	301	(c)	335	(b)
234	(a)	268	(b)	302	(b)	336	(c)
235	(c)	269	(d)	303	(b)	337	(c)
236	(d)	270	(b)	304	(b)	338	(c)
237	(b)	271	(b)	305	(b)	339	(b)
238	(a)	272	(a)	306	(a)	340	(c)
239	(a)	273	(b)	307	(b)	341	(b)
240	(c)	274	(c)	308	(b)	342	(c)
241	(d)	275	(d)	309	(a)	343	(c)
242	(d)	276	(c)	310	(c)	344	(b)
243	(c)	277	(c)	311	(d)	345	(c)
244	(a)	278	(a)	312	(c)	346	(c)
245	(c)	279	(c)	313	(a)	347	(c)
246	(a)	280	(a)	314	(a)	348	(b)
247	(a)	281	(c)	315	(a)	349	(a)
248	(a)	282	(a)	316	(c)		
249	(c)	283	(a)	317	(d)		
250	(c)	284	(a)	318	(d)		

Fill in the Blanks

1. Ocelli
2. Independent eye
3. Ocular sclerites
4. Clavola
5. Mouth parts
6. Hypopharynx (tongue)
7. Labrum
8. Mouth cavity
9. Sub-mentum
10. Epipharynx
11. Bite and grind
12. Labium
13. Left mandible
14. Mouth cone
15. Lapping tongue
16. Lick the nectar
17. Haustellum
18. Apterous
19. Deciduous
20. Thoracic flight muscles
21. Veins
22. Classification
23. Comsock - Needham system
24. Cross veins
25. Furcula or furca
26. Naiads
27. First eight
28. 1st abdominal segment
29. Auditory nerve
30. Physogastry

31. **A male copulatory organ**
32. **Styli**
33. **Angles**
34. **Basitarsus**

Matching

1. a - iv, b - ii, c - i, d - iii
2. a - ii, b - i, c - iv, d - iii
3. a - ii, b - iv, c - i, d - iii
4. a - iii, b - iv, c - ii, d - i
5. a - iii, b - ii, c - i, d - iv
6. a - i, b - iii, c - ii, d - iv
7. a - iii, b - ii, c - i, d - iv
8. a - i, b - iv, c - iii, d - ii
9. a - i, b - ii, c - iii, d - iv
10. a - ii, b - i, c - iv, d - iii

Chapter 3

Embryology, Internal Anatomy and Physiology

Multiple Choice Questions

1. **A series of cell divisions and changes helping in the production of mature ova or egg cells is called**

   (*a*) Spermatogenesis (*b*) Spermiogenesis

   (*c*) Oogenesis (*d*) None of these

2. **The formation of spermatozoa in the testis is known as**

   (*a*) Spermatogenesis (*b*) Spermiogenesis

   (*c*) Oogenesis (*d*) None of these

3. **Which of the following is a function of nephrocytes?**

   (*a*) Storage excretion (*b*) Protein and lipid metabolism

   (*c*) Control of heart beat (*d*) All of these

4. **The amount of uric acid in excreta of insects is about**

   (*a*) 20 per cent (*b*) 50 per cent

   (*c*) 75 per cent (*d*) 85 per cent

5. **Spermatogenesis is normally involves the successive formation of**

   (*a*) Spermatogonia - Secondary spermatocytes - Primary spermatocytes - Spermatids - Spermatozoa

   (*b*) Spermatogonia - Primary spermatocytes - Spermatids - Secondary spermatocytes - Spermatozoa (*c*) Spermatogonia - Spermatids - Secondary spermatocytes - Primary spermatocytes - Spermatozoa

(*d*) Spermatogonia - Primary spermatocytes - Secondary spermatocytes - Spermatids - Spermatozoa

6. **Oogenesis is normally involves the successive formation of**

(*a*) Oogonium - Primary oocytes - Oocyte - Secondary oocytes - Ovam

(*b*) Oogonium - Secondary oocytes - Primary oocytes - Oocyte - Ovam

(*c*) Oogonium - Oocyte - Primary oocytes - Secondary oocytes - Ovam

(*d*) Oogonium - Primary oocytes - Secondary oocytes - Oocyte - Ovam

7. **Excretion of uric acid by insects is a physiological activity mainly because**

(*a*) To reduce water loss (*b*) It is harmless

(*c*) It crystallizes easily (*d*) All of these

8. **Excretory product in red cotton bug is**

(*a*) Uric acid (*b*) Allantoin

(*c*) Ammonia (*d*) Urea

9. **The non-motile female gametes are known as**

(*a*) Pupa (*b*) Sperm

(*c*) Egg (*d*) None of these

10. **Multicellular structure developed from zygote, giving rise to new off spring inside female tissue or egg membrane known as**

(*a*) Egg (*b*) Embryo

(*c*) Ovam (*d*) Oocyte

11. **Fusion of two heterogametes (*i.e.* male and female) is known as**

(*a*) Pollination (*b*) Fertilization

(*c*) Copulation (*d*) Reproduction

12. **Reproduction without fertilization is known as**

(*a*) Parthenogenesis (*b*) Parthenocarpy

(*c*) Pedogenesis (*d*) All of these

13. **The branch of biology which deals with the growth and development of an embryo of an organism, commencing with the union of male and female gametes is known as**

(*a*) Embryology (*b*) Entomology

(*c*) Zoology (*d*) Origin biology

14. Insect eggs are covered by two shells, the tough outer shell is called
(a) Chorion
(b) Vitelline membrane
(c) Micropyle
(d) Spermatheca

15. Which of the following is associated with storage excretion in insect?
(a) Goblet cells
(b) Fat bodies
(c) Malphigian tubules
(d) None of these

16. Major waste product of aquatic insects is
(a) Uric acid
(b) Hydrogen sulphide
(c) Ammonia
(d) Urea

17. Insect eggs are covered by two shells, the thinner inner shell is called
(a) Chorion
(b) Vitelline membrane
(c) Micropyle
(d) Spermatheca

18. The opening at the chorion for the entrance of sperm is known as
(a) Chorion
(b) Vitelline membrane
(c) Micropyle
(d) Spermatheca

19. The sperm of most of insects remain alive for months to years in female body within the
(a) Chorion
(b) Vitelline membrane
(c) Micropyle
(d) Spermatheca

20. Embryonic development of insects takes place inside the
(a) Egg
(b) Cocoon
(c) Pupa
(d) Womb

21. Luciferin is produced by
(a) Labial glands
(b) Malphigian tubules
(c) Head glands
(d) Hypopharyngeal gland

22. Insect nervous system is
(a) Chlorigenic
(b) Andregenic
(c) Both (a) & (b)
(d) None of these

23. Embryonic development is started immediately after__________of the egg.
(a) Formation
(b) Fertilization
(c) Laying
(d) Rupture

24. The cell formed by the union of the male and female gamete is known as
(a) Zygote
(b) Egg
(c) Embryo
(d) Gene

25. Logically Zygote is a

(*a*) Unfertilized ovum (*b*) Fertilized ovum

(*c*) Distorted ovum (*d*) None of these

26. Single zygote divided into__________stage is known as blastomeres.

(*a*) Two cell (*b*) Four cell

(*c*) Eight cell (*d*) Twelve cell

27. In spittle bugs (Cercopids) the sticky substance is produced by

(*a*) Labial glands (*b*) Malphigian tubules

(*c*) Head glands (*d*) None of these

28. The neurons which carry impulses from the central nervous system to sense organs are called

(*a*) Sensory (*b*) Afferent

(*c*) Motor (*d*) Inter neurons

29. Energids form a single cell layer at the periphery of the egg is known as

(*a*) Vitelline membrane (*b*) Blastoderm

(*c*) Endoderm (*d*) None of these

30. Blastoderm forms a thin layer of cuboid cells, they become columner and thicker in the ventral region is known as Embryonic primordia or

(*a*) Gram band (*b*) Germ band

(*c*) Germcell (*d*) None of these

31. The germband develops future embryo, the rest blastoderm remains as extra-embryonic membrane called

(*a*) Cerci (*b*) Serosa

(*c*) Stratum (*d*) Crochet

32. The process by which the mesoderm and endoderm are invaginated within the ectoderm is known as

(*a*) Katatrepsis (*b*) Blastokinesis

(*c*) Gastrulation (*d*) Blastulation

33. The main excretory product of insects is

(*a*) Urea (*b*) Ammonia

(*c*) Uric acid (*d*) None of these

34. Neurocyte is

(*a*) Body of neuron (*b*) Mass of axon

(*c*) Mass of dendrites (*d*) None of these

35. The movement of embryo within the egg is known as

(*a*) Gastrulation (*b*) Blastokinesis
(*c*) Katatrepsis (*d*) Blastulation

36. The movement of embryo from ventral to dorsal surface is known as

(*a*) Anatrepsis (*b*) Blastokinesis
(*c*) Katatrepsis (*d*) Blastulation

37. The movement of embryo taking place from the posterior to anterior pole of egg is known as

(*a*) Anatrepsis (*b*) Blastokinesis
(*c*) Katatrepsis (*d*) Blastulation

38. The insect body wall, tracheal system, nervous system, malpighian tubules, foregut and hindgut are originated from

(*a*) Ectoderm (*b*) Mesoderm
(*c*) Endoderm (*d*) None of these

39. The specialized hydrofuge hairs which hold a permanent thin film of air on the outside of the body of aquatic insects is

(*a*) Aquatron (*b*) Mesotron
(*c*) Plastron (*d*) None of these

40. Neuropile is

(*a*) Body of neuron (*b*) Mass of axon
(*c*) Mass of dendrites (*d*) Glial cells

41. Muscular system, heart and gonads of insects are originated from

(*a*) Ectoderm (*b*) Mesoderm
(*c*) Endoderm (*d*) None of these

42. The insect Midgut are originated from

(*a*) Ectoderm (*b*) Mesoderm
(*c*) Endoderm (*d*) None of these

43. Body segments starts in embryonic development and it involves

(*a*) Ectoderm (*b*) Mesoderm
(*c*) Endoderm (*d*) Both (*a*) & (*b*)

44. The process of separation of old cuticle from the epidermis is known as

(*a*) Apolysis (*b*) Ecdysis
(*c*) Exuvae (*d*) Moulting

45. Spiracles are present in

(*a*) Pleura (*b*) Ventral terga

(*c*) Lateral terga (*d*) Sternum

46. Insect nerves are characteristically

(*a*) Monopolar (*b*) Bipolar

(*c*) Both (*a*) & (*b*) (*d*) None of these

47. The process of final rupture of old cuticle and shedding is known as

(*a*) Apolysis (*b*) Ecdysis

(*c*) Exuvae (*d*) Moulting

48. The insect emerges leaving the cast off skin is known as

(*a*) Apolysis (*b*) Ecdysis

(*c*) Exuvae (*d*) Metamorphosis

49. Specialized form of pupa in which the last larval skin is retained as an extra covering is known as

(*a*) Cocoon (*b*) Puparium

(*c*) Pre pupa (*d*) None of these

50. Insect depositing eggs before hatching is known as

(*a*) Viviparity (*b*) Paedogenesis

(*c*) Oviparous (*d*) Ametabola

51. A respiratory system in which only one pair mesothoracic spiracles are functional is called

(*a*) Holopneustic (*b*) Perineustic

(*c*) Hemipneustic (*d*) Propneustic

52. During polarized condition the potential inside the nerve membrane is

(*a*) Negatively charged (*b*) Positively charged

(*c*) Uncharged (*d*) None of these

53. The condition of giving birth to active young rather than laying eggs are known as

(*a*) Viviparity (*b*) Paedogenesis

(*c*) Oviparous (*d*) Ametabola

54. Production of offspring, by an immature insect is known as

(*a*) Viviparity (*b*) Paedogenesis

(*c*) Oviparous (*d*) Ametabola

55. Insects do not undergo metamorphosis is known as

(a) Hemimetabola (b) Holometabolus

(c) Paurometabola (d) Ametabola

56. Insects in which there is no pupal stage and in which metamorphosis involves very little change is known as

(a) Hemimetabola (b) Holometabolus

(c) Paurometabola (d) Ametabola

57. A respiratory system in which one mesothoracic and eight abdominal spiracles are functional is called

(a) Holopneustic (b) Perineustic

(c) Hemipneustic (d) Propneustic

58. In insect the resting potential of nerve axons is about

(a) - 70 mv (b) - 80 mv

(c) + 70 mv (d) + 50 mv

59. The insects have four life stages are called as

(a) Hemimetabola (b) Holometabolus

(c) Paurometabola (d) Ametabola

60. The gradual metamorphosis also known as

(a) Hemimetabola (b) Holometabolus

(c) Paurometabola (d) Ametabola

61. Grasshopper, Plant bugs are which type of insect

(a) Hemimetabolous (b) Holometabolous

(c) Paurometabolous (d) Ametabolous

62. Eggs of cockroach are encapsulated in a case is known as

(a) Egg pod (b) Egg capsule

(c) Ootheca (d) Ovisack

63. Due to flow of Na^+ inside the axon, the inner membrane becomes positively charged and outer membrane becomes negatively charged, this condition is known as

(a) Resting potential (b) Action potential

(c) Falling phase (d) Repolarisation

64. Which of the following is a neuro transmitter in insects

(a) Acetylcholine (b) L-Glutamate

(c) Gamma - amino butyric acid (d) All of these

65. Larvae having well developed thoracic leg but abdominal legs are absent is known as

(*a*) Oligopod (*b*) Scarabaeiform

(*c*) Eucephalous (*d*) Polypod

66. "C" shaped oligopod larvae are known as

(*a*) Oligopod (*b*) Scarabaeiform

(*c*) Eucephalous (*d*) Polypod

67. Larvae having both thoracic leg and abdominal prolegs are known as

(*a*) Oligopod (*b*) Scarabaeiform

(*c*) Eucephalous (*d*) Polypod

68. Larvae with well developed head capsule and functional mouthparts are known as

(*a*) Acephalous (*b*) Scarabaeiform

(*c*) Eucephalous (*d*) Polypod

69. A respiratory system in which all spiracles are non functional is called

(*a*) Holopneustic (*b*) Aneustic

(*c*) Hemipneustic (*d*) Propneustic

70. The type of muscles found in insects

(*a*) Striated (*b*) Non - striated

(*c*) Unstriated (*d*) Smooth

71. Larvae having no head capsule, mouthparts consists of a pair of protrusible curved mouth hooks is known as

(*a*) Acephalous (*b*) Scarabaeiform

(*c*) Eucephalous (*d*) Polypod

72. Maggot of house fly is a__________type of larvae.

(*a*) Acephalous (*b*) Scarabaeiform

(*c*) Eucephalous (*d*) Polypod

73. Various appendages like, antennae, legs, wing pads are glued with pupal body, type of pupa is

(*a*) Obtect (*b*) Exarate

(*c*) Coarctate (*d*) None of these

74. Various appendages like, antennae, legs, wing pads are not glued with pupal body, pupal type is

(*a*) Obtect (*b*) Exarate

(*c*) Coarctate (*d*) None of these

75. A respiratory system in which all ten pairs of spiracles are functional is called

(a) Holopneustic (b) Perineustic
(c) Hemipneustic (d) Propneustic

76. The unit of muscle between two Z lines is called

(a) Sarcolemma (b) Sarcoplasm
(c) Sarcomere (d) None of these

77. Anterior invagination of ectoderm forms

(a) Stomodeum or foregut (b) Mesenteron or midgut
(c) Proctodeum or hindgut (d) None of these

78. Preoral cavity between hypopharynx and salivary duct is known as

(a) Salivarium (b) Cibarium
(c) Stomodeum (d) None of these

79. Preoral cavity between epipharynx and hypopharynx is known as

(a) Salivarium (b) Cibarium
(c) Stomodeum (d) None of these

80. In sucking insects pharynx act as a

(a) Pollen separator (b) Sucking pump
(c) Food reservoir (d) Food grinder

81. Apneustic condition present in

(a) Chironomid larvae (b) Neuropteran larva
(c) Lepidopteran larvae (d) None of these

82. Actin and Myosin are the proteins present in

(a) Muscles (b) Spermatozoa
(c) Malphigian tubules (d) All of these

83. Crop is the dilated distal part of oesophagous and act as a

(a) Pollen separator (b) Sucking pump
(c) Food reservoir (d) Food grinder

84. The musculated posterior part of foregut is known as

(a) Oesophagous (b) Crop
(c) Gizzard or Proventiculus (d) None of these

85. Food flow from foregut to midgut is regulated through

(a) Oesophageal valve or cardial valve (b) Crop
(c) Gizzard or Proventiculus (d) None of these

86. Gizzard is absent in

(*a*) Leaf cutter (*b*) Bark eater

(*c*) Sap feeders (*d*) Blood sucker

87. Change in volume of tracheal system

(*a*) Ventillation (*b*) Tidal flow

(*c*) Dialysis (*d*) Expansion

88. The sac for storage of sperms in male insect is called

(*a*) Spermatheca (*b*) Seminal vesicle

(*c*) Median oviduct (*d*) Genital chamber

89. Peritrophic membrane is the internal lining of__________present in solid feeders and absent in sap feeders.

(*a*) Foregut (*b*) Midgut

(*c*) Hindgut (*d*) None of these

90. Midgut opens into hindgut through__________which regulate food flow.

(*a*) Oesophageal valve or cardial valve

(*b*) Pyloric valve or proctodeal valve

(*c*) Gizzard or Proventiculus

(*d*) None of these

91. Hindgut is the posterior invagination of

(*a*) Endoderm (*b*) Mesoderm

(*c*) Ectoderm (*d*) None of these

92. Absorption of water, salt and other useful substances from the faeces and urine are the main functions of

(*a*) Foregut (*b*) Midgut

(*c*) Gizzard (*d*) Hindgut

93. Taenidea does not come out at the time of moulting from

(*a*) Spiracles (*b*) Trachea

(*c*) Tracheoles (*d*) Peritreme

94. The function of spermatheca is

(*a*) Receive and store spermatozoa (*b*) Producing of spermatozoa

(*c*) Accumulation of spermatozoa (*d*) None of these

95. Ileum, colon and rectum are three regions of

(*a*) Hindgut (*b*) Foregut

(*c*) Midgut (*d*) Gizzard

96. The liquid faeces of homopteran bugs with soluble sugars and amino acids is known as

(a) Nectar
(b) Honey dew
(c) Dew drop
(d) All of these

97. Carbohydrate in insect diet is essential for

(a) Supply of energy
(b) Synthesis of protein
(c) Synthesized sterol
(d) None of these

98. In insect diet free amino acid required for

(a) Supply of energy
(b) Synthesis of protein
(c) Synthesized sterol
(d) None of these

99. The type of haemocytes which are responsible for plasmatic coagulation

(a) Prohaemocytes
(b) Plasmatocytes
(c) Cystocytes
(d) Granular haemocytes

100. Ovarioles in which eggs alternate with nutritive cells

(a) Polytrophic
(b) Panoistic
(c) Autotrophic
(d) Telotrophic

101. Insect does not synthesized sterol so they required supply of__________in their diet.

(a) Phytosterol
(b) Free amino acid
(c) Glucose
(d) Sucrose

102. Trihalose is a

(a) Monosaccharide
(b) Disaccharide
(c) Trisaccharide
(d) Polysaccharide

103. The excretion process helps the insect to maintain

(a) Salt balance
(b) Water balance
(c) Physiological homeostasis
(d) All of these

104. The chief excretory organs of insects; narrow glandular tubes opening into the alimentary canal near the commencement of the hindgut is known as

(a) Green gland
(b) Malpighian tubules
(c) Hepatic caecae
(d) None of these

105. The type of haemocytes which are responsible for the production of other type of haemocytes

(a) Prohaemocytes
(b) Plasmatocytes
(c) Cystocytes
(d) Granular haemocytes

106. The zone where egg yolks are produced is known as__________.

(*a*) Zone of growth (*b*) Germarium

(*c*) Vitellarium (*d*) Zone of maturation

107. Malpighian tubules are absent in

(*a*) Spring tails (*b*) Aphids

(*c*) Cockroach (*d*) Both (*a*) & (*b*)

108. Terrestrial insects excrete waste products as uric acid or certain of its salts called urates which are water insoluble, this type of excretion is known as

(*a*) Uricotelism (*b*) Ammonotelism

(*c*) Uritelism (*d*) Uriotelism

109. In aquatic insects ammonia is the excretory product, which is freely soluble in water, and more water require for excretion and this type of excretion is known as

(*a*) Uricotelism (*b*) Ammonotelism

(*c*) Ammonialism (*d*) Ammonelism

110. Giving white colour to red cotton bug__________is stored in the body wall.

(*a*) Nitrogen (*b*) Uric acid

(*c*) Amino acid (*d*) Ammonia

111. In insects the heart beat is controlled by

(*a*) Myogenic (*b*) Neurogenic

(*c*) Both (*a*) & (*b*) (*d*) None of these

112. Giving birth of active young, which undergo growth and development inside the mother

(*a*) Viviparous (*b*) Oviparous

(*c*) Ovoviviparous (*d*) None of these

113. The excretory waste materials are retained within the body in different sites is known as

(*a*) Internal excretion (*b*) Storage excretion

(*c*) External excretion (*d*) Inner excretion

114. The external opening of the respiratory system of an insect is known as

(*a*) Spiracle (*b*) Tracheae

(*c*) Taenidia (*d*) Gills

115. The breathing tubes of insect, opening by spiracles situated along the side of the body is known as

(a) Spiracle (b) Tracheae
(c) Taenidia (d) Gills

116. Minute branching tubes leading from the tracheae of insect is known as__________they penetrating to all parts of the body.

(a) Spiracle (b) Tracheoles
(c) Taenidia (d) Gills

117. Characteristic feature of insect haemolymph is

(a) More free amino acids (b) More Trehalose
(c) Both (a) & (b) (d) None of the above

118. Producing active young immediately following the hatching of the egg within the female is known as

(a) Viviparous (b) Oviparous
(c) Ovoviviparous (d) Parthenogenesis

119. The ringed appearance of the tracheae is due to spiral ridges, is known as

(a) Spiracle (b) Tracheoles
(c) Taenidia (d) Gills

120. The main circulatory pump of insect is the pulsetile

(a) Ventral vessel (b) Central vessel
(c) Dorsal vessel (d) Lateral vessel

121. The anterior part of the dorsal vessel is called

(a) Aorta (b) Ostia
(c) Auricle (d) Heart

122. The posterior part of the dorsal vessel is called

(a) Aorta (b) Ostia
(c) Auricle (d) Heart

123. The expanded condition of the heart after diastole is called

(a) Epistatis (b) Distolasis
(c) Mesostatis (d) Diastasis

124. Reproduction by the development of an egg that has not been fertilized with sperm

(a) Parthenogenesis (b) Viviparous
(c) Ovoviviparous (d) Parthenocarpic

125. Openings by which haemolymph flows into the heart from the pericardial sinus in insect is known as

(*a*) Aorta (*b*) Ostia

(*c*) Auricle (*d*) Ventricle

126. Insect Haemolymph is a watery fluid containing

(*a*) Ions (*b*) Molecules

(*c*) Cells (*d*) All of these

127. Exudation of haemolymph through slit, pore etc. is known as

(*a*) Reflex bleeding (*b*) Coagulation bleeding

(*c*) Impulse bleeding (*d*) Haemorrhage

128. Reflex bleeding of Aphid, may natural enemy found in

(*a*) Attracts (*b*) Repels

(*c*) Invites (*d*) None of these

129. Salivary glands of insects are also called

(*a*) Mandibular glands (*b*) Maxillary glands

(*c*) Labial glands (*d*) Pharyngeal glands

130. When embryonic development is completed within the body of the female, it is called

(*a*) Viviparity (*b*) Oviparity

(*c*) Ovoviviparity (*d*) Parthenogenesis

131. Abnormal individual with secondary sexual characters of both male and female is known as

(*a*) Masculinemorph (*b*) Femininemorph

(*c*) Gynandromorph (*d*) Reciprocalmorph

132. Male and female gonads are present in one organism is known as

(*a*) Hermaphrodite (*b*) Gynandromorph

(*c*) Reciprocalmorph (*d*) All of these

133. Specialized nutritive cells for nourishment of oocytes is known as

(*a*) Nursocytes (*b*) Cellocytes

(*c*) Trophocytes (*d*) None of these

134. In which oocytes grow by deposition of yolk in a process, is known as

(*a*) Oogenesis (*b*) Vitellogenesis

(*c*) Choriosis (*d*) All of these

135. Filter chamber is present in

(a) Homoptera (b) Heteroptera
(c) Thysanoptera (d) Diptera

136. Haemocoelous viviparity is prominent in the insect order

(a) Lepidoptera (b) Strepsiptera
(c) Diptera (d) Coleoptera

137. Venom of bee is bee is produced by accessory glands of the female

(a) Excretory system (b) Respiratory system
(c) Reproductive system (d) Salivary system

138. Pheromone glands are situated on

(a) Abdominal terminalia (b) Thoracic terminalia
(c) Thoracic ganglia (d) None of these

139. Corpora cardiaca controls the heartbeat and regulate the level of__________in haemolymph.

(a) Uric acid (b) Triglyceride
(c) Trehalose (d) All of these

140. A fusion of corpora cardiaca, corpora allata, prothoracic gland and hypocerebral ganglia, which secrets hormone for hardening of puparium are found in maggots of dipteran fly is known as

(a) Johnston's organ (b) Weismann's ring
(c) Johnston's ring (d) Chordotonal organ

141. Extra oral digestion is found in

(a) Leaf chewing insects (b) Sap sucking insects
(c) Bark eating caterpillar (d) All of these

142. The type of parthenogenesis in which only females are produced

(a) Arrhenotoky (b) Thelytoky
(c) Amphitoky (d) None of these

143. Which part of the nervous system receive stimuli

(a) Dendrites (b) Axon
(c) Synapse (d) None of these

144. The information transmit to another neuron or muscle by

(a) Dendrites (b) Axon
(c) Synapse (d) None of these

145. The terminal arborization of axon is known as

(*a*) Dendrites (*b*) Tentacle
(*c*) Synapse (*d*) None of these

146. The important neurotransmitter of insects is

(*a*) Acetylecholine (*b*) Catecholamines
(*c*) Dopamine (*d*) Both (*a*) & (*b*)

147. Generally gut pH of insect is varies between

(*a*) 5.5 - 6.5 (*b*) 6.5 - 7.5
(*c*) 7.5 - 8.5 (*d*) 8 - 10

148. Which of the following sequence is correct

(*a*) Spermatogonia - Spermatocytes - Spermatids - Spermatozoa
(*b*) Spermatogonia - Spermatocytes - Spermatozoa - Spermatids
(*c*) Spermatocytes - Spermatogonia - Spermatids - Spermatozoa
(*d*) Spermatids - Spermatogonia - Spermatocytes - Spermatozoa

149. A distinct swelling of nervous tissue in which many inter connection between nerve ending occurs, is known as

(*a*) Synaps (*b*) Ganglion
(*c*) Nervous linkage (*d*) None of these

150. Brain of insect consists of

(*a*) Protocerebrum (*b*) Deutocerebrum
(*c*) Tritotocerebrum (*d*) All of these

151. The specialized sensory organs that receive vibrations are subcuticular mechano receptors are known as

(*a*) Johnston's organ (*b*) Weismann's ring
(*c*) Johnston's ring (*d*) Chordotonal organ

152. A collection of chordotonal receptors present on the second segment (pedicel) of antennae is known

(*a*) Johnston's organ (*b*) Weismann's ring
(*c*) Johnston's ring (*d*) Chordotonal organ

153. The parts involved in the formation of filter chamber are anterior ends of

(*a*) Mid and Hind gut (*b*) Fore and Mid gut
(*c*) Fore and Hind gut (*d*) None of these

154. Cyclic parthenogenesis is seen in

(*a*) Aphids (*b*) Thrips
(*c*) Mealy bugs (*d*) Grasshoppers

155. The only visual organs of holometabolous larvae is known as

(*a*) Stemmata or lateral ocelli (*b*) Dorsal ocelli

(*c*) Compound eye (*d*) All of these

156. Visual organs of nymph is known as__________it perceive light to maintain diurnal rhythm and not involved in image perception.

(*a*) Stemmata or lateral ocelli (*b*) Dorsal ocelli

(*c*) Compound eye (*d*) All of these

157. The compound eye of insect is based on many individual unit is called

(*a*) Retina (*b*) Ocelli

(*c*) Ommatidia (*d*) None of these

158. The secondary pigment cells of an omatidia secretes rod shaped rhabdom which contains light sensitive pigment are called

(*a*) Retinine (*b*) Omatidin

(*c*) Rhodopsin (*d*) Adopsin

159. Alimentary canal is longer in__________insects.

(*a*) Carnivorous (*b*) Phytophagous

(*c*) Parasitic (*d*) Scavengers

160. The cleavage cells which remain in the yolk or return to it after reaching the periphery of the egg are

(*a*) Vitellophages (*b*) Endocytes

(*c*) Plasmocytes (*d*) Oocytes

161. In diurnal insects ommatidium are of apposition type, presence of primary pigment cell, which produce

(*a*) Mosaic image (*b*) Bright image

(*c*) Dark image (*d*) None of these

162. In nocturnal and crepuscular insects which of the following is correct

(*a*) Ommatidium are of super position type

(*b*) Primary pigment cells are absent

(*c*) Images are indistinct, bright and blurred

(*d*) All of these

163. On structural basis Neuron with a single axon, the type is

(*a*) Monopolar (*b*) Bipolar

(*c*) Multipolar (*d*) Nonpolar

164. On structural basis Neuron with a proximal axon and a long distal dendrite, the type is

(*a*) Monopolar (*b*) Bipolar

(*c*) Multipolar (*d*) Nonpolar

165. Proctodaeum is also called as

(*a*) Foregut (*b*) Midgut

(*c*) Hindgut (*d*) All of these

166. The process by which mesoderm and endoderm are invaginated within the ectoderm is called

(*a*) Blastrulation (*b*) Gastrulation

(*c*) Blastokinesis (*d*) Vitellogenesis

167. On structural basis Neuron with a proximal axon and many distal dendrites, the type is

(*a*) Monopolar (*b*) Bipolar

(*c*) Multipolar (*d*) Nonpolar

168. In which ovarioles lacks of trophocytes, the type is known as

(*a*) Paniostic ovariole (*b*) Polytrophic ovariole

(*c*) Telotrophic or Acrotrophic ovariole (*d*) None of these

169. In which ovarioles trophocytes are confined the type is known as

(*a*) Paniostic ovariole (*b*) Polytrophic ovariole

(*c*) Telotrophic or Acrotrophic ovariole (*d*) None of these

170. In which ovarioles trophocytes are connected to each oocytes the type is known as

(*a*) Paniostic ovariole (*b*) Polytrophic ovariole

(*c*) Telotrophic or Acrotrophic ovariole (*d*) None of these

171. The expanded portion of the foregut just posterior to the oesophagus is

(*a*) Proventriculus (*b*) Crop

(*c*) Small intestine (*d*) Ventriculus

172. The one - cell thick layer of cells that surrounds the yolk of an egg are

(*a*) Ectoderm (*b*) Blastoderm

(*c*) Endoderm (*d*) Mesoderm

173. In which insect Haemocoelous sperm transfer are found

(*a*) Silver fish (*b*) Bed bug

(*c*) Cockroach (*d*) None of these

174. The neuron which conducts impulse from sense organs to CNS is known as

(*a*) Sensory neuron (*b*) Motor neuron
(*c*) Inter neuron (*d*) None of these

175. The neuron which conducts impulse from CNS to effector organ is known as

(*a*) Sensory neuron (*b*) Motor neuron
(*c*) Inter neuron (*d*) None of these

176. The sac for storage of sperms in female insect is called

(*a*) Ootheca (*b*) Median oviduct
(*c*) Genital chamber (*d*) Spermatheca

177. Proventriculus or Gizzard is a part of

(*a*) Foregut (*b*) Midgut
(*c*) Hindgut (*d*) Filter chamber

178. Which of the following is not ectodermal origin?

(*a*) Foregut and Hindgut (*b*) Nervous system
(*c*) Respiratory system (*d*) Fat body

179. Ecdysial cleavage line helps in

(*a*) Apolysis (*b*) Ecdysis
(*c*) Both (*a*) & (*b*) (*d*) None of these

180. Detachment of old cuticle from epidermis is

(*a*) Ecdysis (*b*) Apolysis
(*c*) Moulting (*d*) None of these

181. Time stage between two subsequent moulting

(*a*) Prepupa (*b*) Stadium
(*c*) Instar (*d*) None of these

182. Differentiation of the insect body into distinct functional regions

(*a*) Oogenesis (*b*) Tagmosis
(*c*) Morphogenesis (*d*) Spermatogenesis

183. Foregut of insect is lined by Cuticle is called as

(*a*) Intima (*b*) Peritrophic membrane
(*c*) Epithelium (*d*) Taenidia

184. Which of the following is mechanoreceptors?

(*a*) Trichoid sensilla (*b*) Campaniform sensilla
(*c*) Chordotonal organs (*d*) All of these

185. Foregut and hindgut of insect are lined by

(*a*) Epidermis (*b*) Basement membrane

(*c*) Cuticle (*d*) None of these

186. Which of the following is ectodermal in origin?

(*a*) Foregut (*b*) Midgut

(*c*) Hindgut (*d*) Both (*a*) & (*c*) and c

187. Plate like gizzard is found in

(*a*) Cockroach (*b*) Honey bee

(*c*) Flea (*d*) Butterfly

188. Which of the following is true about peritrophic membrane?

(*a*) Present in sap feeders and absent in solid feeder

(*b*) Present in solid feeders and absent in sap feeder

(*c*) Present in both sap feeders and solid feeder

(*d*) All of these

189. Which of the following is a uni-cellular integumentary outgrowth?

(*a*) Scales (*b*) Setae

(*c*) Bristles (*d*) All of these

190. In Acrididae, the tympanal organs are present on the

(*a*) 1st abdominal tergum (*b*) 2nd abdominal tergum

(*c*) 1st abdominal sternum (*d*) 2nd abdominal sternum

191. Proventriculus is associated with

(*a*) Foregut (*b*) Midgut

(*c*) Hindgut (*d*) None of these

192. Filter chamber present in

(*a*) Lepidopteran insect (*b*) Coleopteran insect

(*c*) Hemipteran insect (*d*) Collembolan insect

193. In honey bee saliva contains

(*a*) Malase (*b*) Invertase

(*c*) Protease (*d*) Lipase

194. Digestive enzyme of bird lice is

(*a*) Cellulase (*b*) Collagenase

(*c*) Keratinase (*d*) Both (*a*) & (*b*)

195. Immobile multicellular integumentary outgrowth is

(a) Spur (b) Spine
(c) Claw (d) None of these

196. In Tettigonidae and Grylloidea a pair of tympanal organs are present on the

(a) Fore tibia (b) Mid tibia
(c) Hind tibia (d) None of these

197. Digestive enzyme of meat eating maggots is

(a) Cellulase (b) Collagenase
(c) Keratinase (d) Both *(a)* & *(b)*

198. In wax moth which of the following is helps in wax digestion

(a) Bacteria (b) Fungi
(c) Protozoa (d) Virus

199. Cryptonephry is found in

(a) Larval Lepidoptera (b) Adult Coleoptera
(c) Larval Symphyta (d) All of these

200. Malpighian tubules help in silk production in

(a) Larval Collembola (b) Larval Neuroptera
(c) Adult Thysanura (d) Adult Protura

201. Mobile multicellular integumentary outgrowth is

(a) Spur (b) Spine
(c) Claw (d) None of these

202. A highly specialized chordotonal organs located in the pedicel of the antenna are

(a) Johnston's organ (b) Dufaur's gland
(c) Stobbes gland (d) None of these

203. Uric acid is stored in the wing scales giving white colour in which of the following insects

(a) White stem borer moth (b) Woolly aphid
(c) Pierid butterfly (d) All of these

204. Meconium is the waste product of

(a) Larval metabolism (b) Pupal metabolism
(c) Adult metabolism (d) Egg metabolism

205. Propneustic type of respiratory system present in some pupa of

(a) Caterpillar (b) Honey bee
(c) Wriggler (d) Puparium

206. The diameter of tracheoles are

(a) 1 µm (b) < 1 µm
(c) > 1 µm (d) 2 µm

207. Socket producing cell is called

(a) Trichogen cell (b) Tormogen cells
(c) Trophocyte (d) Oenocyte

208. Chemoreceptors are present in

(a) Tarsi (b) Maxillae
(c) Mandibles (d) Both (b) & (c)

209. Taenidia and peritreme are associated with the system of

(a) Respiration (b) Circulation
(c) Digestion (d) Reproduction

210. Sexual mosaic is associated with

(a) Cottony cushion scale (b) Aphid
(c) Mutant Drosophila (d) Cockroach

211. External sperm transfer is an important character of

(a) Bed bug (b) Silver fish
(c) Spring tail (d) All of these

212. The concentration of Juvenile hormone__________as the larvae becomes pupae.

(a) Remain same (b) Decreases
(c) Increases (d) None of these

213. Which of the following function as chemoreceptors in insects?

(a) Sensilla trichoidea (b) Sensilla basiconia
(c) Sensilla coeloconia (d) All of these

214. Production of only female is called

(a) Thelytoky (b) Arrhenotoky
(c) Deuterotoky (d) None of these

215. Gradual metamorphosis is seen in

(a) Cockroach (b) Grasshopper
(c) Bugs (d) All of these

216. Scarabaeform larvae posses the shape

(a) U (b) C

(c) O (d) I

217. Larvae of housefly and robber fly are

(a) Eucephalous (b) Hemicephalous

(c) Acephalous (d) None of these

218. Juvenile hormone is secreted by

(a) Prothoracic gland (b) Corpora allata

(c) Corpora cardiaca (d) Neurosecretory cells

219. Biting and chewing insects have contact chemoreceptors on all the mouth parts except the

(a) Labrum (b) Mandibles

(c) Maxillae (d) Labium

220. Earthen cocoon which is made up of soil and saliva done by

(a) Woolly borer

(b) Coconut black headed caterpillar

(c) Housefly

(d) Gram pod borer

221. Which of the following is an abdominal appendage?

(a) Waist (b) Apophysis

(c) Pronotum (d) Gonopore

222. Moulting fluid is able to digest

(a) Cement layer (b) Endocuticle

(c) Exocuticle (d) None of these

223. Which part of moult does not occur during apolysis?

(a) Formation of subcuticular space

(b) Detachment of cuticle from epidermis

(c) Secretion of moulting gel

(d) Removal of old cuticle

224. Ecdysone or moulting hormone is secreted by

(a) Prothoracic gland (b) Corpora allata

(c) Corpora cardiac (d) Neuro secretary cells

225. Which of the following functions as photoreceptors in insect?

(*a*) Dorsal ocelli (*b*) Lateral ocelli or stemmata

(*c*) Compound eyes (*d*) All of these

226. Which part of insect regulate the Fe and H_2O balance in blood and control the internal environment inside the insect?

(*a*) Alimentary system (*b*) Malphigian tubule

(*c*) Reproductive system (*d*) Nervous system

227. Colon is a part of

(*a*) Foregut (*b*) Midgut

(*c*) Hindgut (*d*) Mouth parts

228. The hind brain is also called

(*a*) Protocerebrum (*b*) Duetocerebrum

(*c*) Tritocerebrum (*d*) None of these

229. The respiratory system in which the first and last pair of spiracles is functional is called

(*a*) Metapneustic (*b*) Amphipneustic

(*c*) Peripneustic (*d*) Proppneustic

230. Which of the following is a precursor for ecdysone synthesis?

(*a*) Ergosterol (*b*) Erosteroid

(*c*) Methionine (*d*) Cholesterol

231. Which of the following is a part of the dorsal ocelli?

(*a*) Cornea and corneagen layer (*b*) Retina

(*c*) Pigment cells (*d*) All of these

232. The blood cells which ingest dead cell are known as

(*a*) Phagocytes (*b*) Oenocytes

(*c*) Proleucocytes (*d*) Granulocytes

233. Nephrocytes are related with

(*a*) Respiration (*b*) Excretion

(*c*) Reproduction (*d*) Digestion

234. Complete metamorphosis is found in

(*a*) Silver fish (*b*) Grasshopper

(*c*) Cockroach (*d*) Butterfly

235. The process of ecdysis in insect is related with

(a) Respiration (b) Excretion
(c) Reproduction (d) Growth

236. The lipoprotein layer of the epicuticle is

(a) Cuticulin (b) Polyphenol layer
(c) Cement layer (d) Wax layer

237. A transparent body that lies beneath the cornea in an ommatidium

(a) Ommatophore (b) Crystalline cone
(c) Cornea (d) Retina

238. The polyembryony founds in

(a) Some Orthoptera (b) Some aquatic Odonata
(c) Some parasitic Hymenoptera (d) None of these

239. The metamorphosis found in cockroach is

(a) Polymetabola (b) Paurometabola
(c) Hemimetabola (d) Holometabola

240. Special type of reproduction found in aphids is known as

(a) Vivipary (b) Paedogenesis
(c) Polyembryony (d) Parthenogenesis

241. In insect total number of stage found in complete metamorphosis is

(a) 3 (b) 4
(c) 5 (d) 6

242. The noncellular membrane underlying the epidermal cells of the integument

(a) Basement Membrane (b) Epidermis
(c) Cement layer (d) Wax layer

243. Degenerated ocelli of cockroach are known as

(a) Ommatidium (b) Ocellium
(c) Stergites (d) Fenestra

244. Malpighian tubules arises from

(a) Foregut (b) Midgut
(c) Hindgut (d) None of these

245. How many spiracles are functional in Grasshopper?

(a) 10 pairs (b) 9 pairs
(c) 11 pairs (d) 12 pairs

246. Complete metamorphosis found in

(*a*) Cricket
(*b*) Beetle
(*c*) Bugs
(*d*) Aphid

247. The potential produced in the dendrite of the sense organ by depolarization is called

(*a*) Resting potential
(*b*) Receptor potential
(*c*) Generator potential
(*d*) Action potential

248. Typical insect integument comprises of

(*a*) Cuticle, Epidermis and Basement membrane
(*b*) Cuticle and Epidermis
(*c*) Cuticle and Basement membrane
(*d*) Epidermis and Basement membrane

249. Supra oesophageal ganglion is also called

(*a*) Brain
(*b*) Abdominal ganglia
(*c*) Thoracic ganglia
(*d*) Peripheral Nervous System

250. The process of post ecsysial hardening and darkening of the cuticle is regulated by

(*a*) Ecdysone
(*b*) PTTH
(*c*) Juvenile hormone
(*d*) Bursicon

251. The PTTH is secreted from

(*a*) Prothoraic gland
(*b*) Neurosecretory cells
(*c*) Corpora allata
(*d*) Corpora cardiaca

252. Brain is also known as__________ganglia.

(*a*) Supra oesophageal
(*b*) Sub oesophageal
(*c*) Thoracic
(*d*) Abdominal

253. Only one generation in a year is called

(*a*) Univoltine
(*b*) Bivoltine
(*c*) Multivoltine
(*d*) None of these

254. Chitin is abundant in

(*a*) Epicuticle
(*b*) Exocuticle
(*c*) Mesocuticle
(*d*) Endocuticle

255. The light producing substance in glow worm fire fly is

(*a*) Lucifernine
(*b*) Luciferase
(*c*) Lucithrin
(*d*) Luciferin

256. Only two generations in a year is called

(a) Univoltine
(b) Bivoltine
(c) Multivoltine
(d) None of these

257. Several generations in a year is called

(a) Univoltine
(b) Bivoltine
(c) Multivoltine
(d) None of these

258. Damaging enzyme of woollen cloth is

(a) Keratinase
(b) Ligases
(c) Aldolase
(d) Kinase

259. Alimentary canal of insects generally is divided into

(a) Not divisible
(b) Two parts
(c) Three parts
(d) Many parts

260. Cement layer is absent in

(a) Grasshopper
(b) Honeybees
(c) Cockroaches
(d) All of these

261. Adult diapause is seen in

(a) *Leptinotarsa*
(b) White grub
(c) *Anthonomus*
(d) All of these

262. Excretion removes

(a) Metabolic waste
(b) Undigested material
(c) Stored material
(d) None of these

263. Salivary gland of insect secrete

(a) Amylase
(b) Maltase
(c) Sucrose
(d) None of these

264. Circulatory system of insect is of

(a) Open type
(b) Close type
(c) Semi open type
(d) None of these

265. Insects have

(a) Lungs and Trachea
(b) Trachea and Tracheole
(c) Lungs and Air Sac
(d) None of these

266. Cuticle forms the inner lining of

(a) Fore and Hind gut
(b) Reproductive tract
(c) Tracheae
(d) All of these

267. Sempers rib is

(*a*) Degenerated ocelli
(*b*) Degenerated compound eye
(*c*) Degenerated trachea in wing
(*d*) Degenerated wings

268. Nerve cord is located

(*a*) Ventrally
(*b*) Dorsally
(*c*) Laterally
(*d*) None of these

269. The endocrine gland secrete

(*a*) Pheromone
(*b*) Hormone
(*c*) Enzyme
(*d*) None of these

270. Insect pheromone secretes from

(*a*) Endocrine gland
(*b*) Exocrine gland
(*c*) Both (*a*) & (*b*)
(*d*) None of these

271. The instar is

(*a*) The actual stage between two moults
(*b*) The time interval between two moults
(*c*) Both (*a*) & (*b*)
(*d*) None of these

272. Sequence of different layers in epicuticle from out side

(*a*) Cement layer - Wax layer - Poly phenol layer - Cuticulin layer
(*b*) Cement layer - Poly phenol layer - Wax layer - Cuticulin layer
(*c*) Wax layer - Cement layer - Poly phenol layer - Cuticulin layer
(*d*) Cement layer - Wax layer - Cuticulin layer - Poly phenol layer

273. Among the following which one is the most noisy insect

(*a*) Crickets
(*b*) Gryllids
(*c*) Male Cicadas
(*d*) Grasshoppers

274. The Stadium is

(*a*) The actual stage between two moults
(*b*) The time interval between two moults
(*c*) Both (*a*) & (*b*)
(*d*) None of these

275. Hepatic cacecae is located at

(*a*) Foregut
(*b*) Midgut
(*c*) Hindgut
(*d*) None of these

276. Intima absent in

(*a*) Foregut (*b*) Midgut

(*c*) Oesophagus (*d*) All of these

277. Alary muscle are associated with

(*a*) Heart (*b*) Trachea

(*c*) Legs (*d*) Wings

278. Chitin is absent in

(*a*) Epicuticle (*b*) Procuticle

(*c*) Mesocuticle (*d*) None of these

279. Carnitine is

(*a*) Insect vitamin (*b*) Insect protein

(*c*) Insecticide (*d*) Insect chitin

280. How many parts of brain are there in a grasshopper?

(*a*) Two (*b*) Three

(*c*) Four (*d*) Five

281. Hemocoel of insects consists of

(*a*) Pericardial sinus (*b*) Perivisceral sinus

(*c*) Perinural sinus (*d*) None of these

282. Metamorphosis includes

(*a*) Changes in larval development (*b*) Changes from egg to adult

(*c*) Changes in nymph (*d*) Changes in behavior of insect

283. The young one of Lepidoptera is known as

(*a*) Maggot (*b*) Caterpillar

(*c*) Grub (*d*) Nymph

284. The young one of Diptera is known as

(*a*) Maggot (*b*) Caterpillar

(*c*) Grub (*d*) Nymph

285. The young one of Coleoptera is known as

(*a*) Maggot (*b*) Caterpillar

(*c*) Grub (*d*) Nymph

286. The young one of Hemiptera is known as

(*a*) Maggot (*b*) Caterpillar

(*c*) Grub (*d*) Nymph

287. Exocuticle and Endocuticle together form

(*a*) Epicuticle (*b*) Procuticle

(*c*) Mesocuticle (*d*) None of these

288. Blister beetles produced an irritating substance

(*a*) Canthamycin (*b*) Canthacaffine

(*c*) Cantharodone (*d*) Cantharidin

289. Reabsorption of water in insects take place in

(*a*) Midgut (*b*) Rectum

(*c*) Hepatic caecea (*d*) Ventriculus

290. The colour of grasshopper blood is

(*a*) Yellow (*b*) Green

(*c*) Colourless (*d*) Blue

291. Plastron respiration occur in

(*a*) Terrestrial insects (*b*) Aquatic insects

(*c*) Soil insects (*d*) None of these

292. Sequence of different layers of insect integument from outside

(*a*) Epicuticle - Exocuticle - Mesocuticle - Endocuticle - Epidermis - Basement Membrane

(*b*) Exocuticle - Epicuticle - Mesocuticle - Endocuticle - Epidermis - Basement Membrane

(*c*) Epicuticle - Exocuticle - Mesocuticle - Epidermis - Endocuticle - Basement Membrane

(*d*) Epicuticle - Exocuticle - Mesocuticle - Endocuticle - Epidermis - Basement Membrane

293. An adhesive pad between the claws of the tarsus

(*a*) Empodium (*b*) Aedeagus

(*c*) Arolium (*d*) Clavus

294. Stage of insects may be fully immotile

(*a*) Egg (*b*) Larvae

(*c*) Pupa (*d*) Adult

295. The main function of peritrophic membrane is

(*a*) Production (*b*) Assimilation

(*c*) Protection (*d*) Absorption

296. Peritrophic membrane is absent in

(a) Orthoptea (b) Hemiptera

(c) Hymenoptera (d) Coleoptera

297. Haemoglobin is found in which insects larvae

(a) Cecidomyid larvae (b) Chironomid larvae

(c) Mycetophilid larvae (d) Psycodid larvae

298. Example of order having pupal stage only in male

(a) Dictyoptera (b) Orthoptera

(c) Isoptera (d) Strepsiptera

299. Berlese's organ found in

(a) Head louse (b) Male mosquito

(c) Female bed bug (d) Grasshopper

300. Glycerophosphate shuttle is found in

(a) Flight muscles (b) Sense organs

(c) Fat bodies (d) None of these

301. Crushing of food in insect is done by

(a) Rectal papilla (b) Oesophagous

(c) Proventriculus (d) Crop

302. Feeding in insects can be checked by the use of

(a) Hormone (b) Pheromone

(c) Feeding deterrent (d) Antibiosis

303. The non cellular part of insect body is

(a) Antenna (b) Cuticle

(c) Legs (d) Cerci

304. Chitin constitutes nearly__________per cent of dry weight of insect.

(a) 15 - 20 per cent (b) 20 - 25 per cent

(c) 25 - 40 per cent (d) 45 - 50 per cent

305. In the formation of cuticle which of the following layer is secreted first

(a) Exocuticle (b) Endocuticle

(c) Wax layer (d) Cuticulin layer

306. Insect having prolegs in larval stage

(a) *Thrips tabaci* (b) *Athalia lugens proxima*

(c) *Dacus cucurbitae* (d) *Holotrichia consanguinea*

307. Which of the following organs are involved in excretion?

(*a*) Malpighian tubules
(*b*) Integument
(*c*) Wall of alimentary canal
(*d*) All of these

308. Which of the following are not haemocytes?

(*a*) Plasmocytes
(*b*) Cytocytes
(*c*) Nephrocytes
(*d*) Coagulocytes

309. The reabsorption of water in insect takes place in

(*a*) Malpighian tubules
(*b*) Hepaticcaecae
(*c*) Rectal papillae
(*d*) Ventriculus

310. Generally pH of insect plasma is acidic and the range near about

(*a*) 5.5
(*b*) 6.7
(*c*) 7.8
(*d*) 8.5

311. Father of insect physiology is

(*a*) Snodgrass
(*b*) Painter
(*c*) Wigglesworth
(*d*) Knipling

312. Alternation of gamic and agamic generations is seen in

(*a*) Aphids
(*b*) Whiteflies
(*c*) Mealy bugs
(*d*) All of these

313. The main circulatory organ in grasshopper is

(*a*) Pericardial sinus
(*b*) Haemoceal
(*c*) Dorsal vessel
(*d*) Perineural sinus

314. The measure of the oxidizing or reducing conditions in a system is known as

(*a*) Positive potential
(*b*) Negative potential
(*c*) Redox potential
(*d*) None of these

Fill in the Blanks

1. The daughter nucleus of the zygote are accompanied by a hollow mass of cytoplasm forming nucleo-cytoplasm unit known as__________.

2. The developmental period after hatching of egg to obtain adult hood is known as__________.

3. The active immature stage of an insect hatched out from egg and under goes complete metamorphosis is known as__________.

4. A non-feeding and relatively inactive stage between the larva and adult in holometabolous insects is known as_________.

5. The sexually matured form or stage of an insect is known as_________.

6. Insect cast off their skin many times during its life time is known as_________.

7. Total period between any two successive moult is called_________.

8. The actual form of insect during a stadium is termed as_________.

9. Specialized form of pupa in which the last larval skin is retained as an extra covering is known as Puparium and the pupal type is_________.

10. The alimentary canal of insect is a long, muscular and tubular structure, differentiated into three regions_________.

11. Midgut of insect is endodermal in origin and also called as_________.

12. Food exchange by two insect individual is known as_________.

13. Trihalose a Disacharide found in haemolymph of insect, used as fuel for_________.

14. Finger like outgrowths found in anterior or posterior ends of midgut is known as_________.

15. Removal of metabolism waste products especially nitrogenous compounds from the body of insect is known as_________.

16. When male and female occur separately is known as_________.

17. An invagination of the body wall around the genital aperture of insect adopted for receiving the intromitter organ of the male is known as_________.

Match the following

1. Match the following nervous system attached with their attached organ

(a) Central Nervous System	i) Brain
(b) Stomatogastic Nervous System	ii) Ventral visceral
(c) Visceral Nervous System	iii) Aorta and foregut
(d) Peripheral Nervous System	iv) Muscels

2. Match different types of parthenogenesis

(a) Cyclic or sporadic	i) Alternation of gamic and agamic population
(b) Arrhenotoky	ii) Produce both male and female

(c)	Thelytoky	iii)	Produce female
(d)	Amphitoky or deuterotoky	iv)	Produce male

3. Match internal cuticle of gizzard is variously modified as in insects

(a)	Cockroach	i)	Plate like
(b)	Honey Bee	ii)	Teeth like
(c)	Aphid	iii)	Spine like
(d)	Flea	iv)	Absent

4. Match accessory functions of malpighin tubules found in insect

(a)	Spittle bug	i)	Spittle secretion
(b)	Spring tail	ii)	Light production
(c)	Bolitophila	iii)	Absent
(d)	Larval neuropteran	iv)	Silk production

5. Match the number and location of functional spiracle

(a)	Holopneustic	i)	2 Thoracic and 8 Abdominal pairs are functional
(b)	Hemipneustic	ii)	Among 10 pairs One or Two non functional
(c)	Peripneustic	iii)	1 Thoracic and 8 Abdominal pairs are functional
(d)	Hypopneustic	iv)	1 Thoracic and 6 Abdominal pairs are functional

6. Match the number and location of functional spiracle

(a)	Amphipneustic	i)	One anterior pair and one posterior pair are functional
(b)	Propneustic	ii)	Only one anterior pair is functional
(c)	Metapneustic	iii)	Only one posterior pair is functional
(d)	Apneustic	iv)	All spiracle closed (Closed tracheal system)

7. Match the following type of respiratory found in insects

(a)	Amphipneustic	i)	Maggot
(b)	Propneustic	ii)	Puparium
(c)	Metapneustic	iii)	Wriggler
(d)	Apneustic	iv)	Naiad of May fly

8. Match the following type of reproduction and insects

(a)	Ovoviviparity	i) Thrips
(b)	Pseudoplacental viviparity	ii) Aphid
(c)	Haemocoelous viviparity	iii) Stepsiptera
(d)	Adenotrophic viviparity	iv) Tsetse fly

9. Match the following insect with their egg laying

(a)	Ootheca	i) Culex mosquito
(b)	Egg pod	ii) Grasshopper
(c)	Egg case	iii) Cockroach
(d)	Egg raft	iv) Mantids

10. Match the following pupal type with their insect

(a)	Chrysalis	i) Butterfly
(b)	Tumbler	ii) Rhinoceros beetle
(c)	Exarate	iii) House fly
(d)	Puparium	iv) Mosquito

11. Match the following defence behaviour found in insects

(a)	Homocromism	i) Cow bug
(b)	Homomorphism	ii) Stick insect
(c)	Homotypism	iii) Preying mantids
(d)	Horney integument	iv) Jewel beetle

Answers

Multiple Choice Questions

1	(c)	21	(b)	41	(b)	61	(c)
2	(a)	22	(a)	42	(c)	62	(c)
3	(d)	23	(b)	43	(d)	63	(b)
4	(d)	24	(a)	44	(a)	64	(d)
5	(d)	25	(b)	45	(c)	65	(a)
6	(d)	26	(a)	46	(a)	66	(b)
7	(d)	27	(b)	47	(b)	67	(d)
8	(b)	28	(c)	48	(c)	68	(c)
9	(c)	29	(b)	49	(b)	69	(b)
10	(b)	30	(b)	50	(c)	70	(a)
11	(b)	31	(b)	51	(d)	71	(a)
12	(a)	32	(c)	52	(a)	72	(a)
13	(a)	33	(c)	53	(a)	73	(a)
14	(a)	34	(a)	54	(b)	74	(b)
15	(b)	35	(b)	55	(d)	75	(a)
16	(c)	36	(c)	56	(a)	76	(c)
17	(b)	37	(a)	57	(b)	77	(a)
18	(c)	38	(a)	58	(a)	78	(a)
19	(d)	39	(c)	59	(b)	79	(b)
20	(a)	40	(b)	60	(c)	80	(b)

81	(a)	115	(b)	149	(b)	183	(a)
82	(a)	116	(b)	150	(d)	184	(d)
83	(c)	117	(c)	151	(d)	185	(c)
84	(c)	118	(c)	152	(a)	186	(d)
85	(a)	119	(c)	153	(a)	187	(b)
86	(c)	120	(c)	154	(a)	188	(b)
87	(a)	121	(a)	155	(a)	189	(d)
88	(b)	122	(d)	156	(b)	190	(a)
89	(b)	123	(d)	157	(c)	191	(a)
90	(b)	124	(a)	158	(c)	192	(c)
91	(c)	125	(b)	159	(b)	193	(b)
92	(d)	126	(d)	160	(a)	194	(c)
93	(c)	127	(a)	161	(a)	195	(b)
94	(a)	128	(b)	162	(d)	196	(a)
95	(a)	129	(c)	163	(a)	197	(b)
96	(b)	130	(a)	164	(b)	198	(a)
97	(a)	131	(c)	165	(c)	199	(d)
98	(b)	132	(a)	166	(b)	200	(b)
99	(c)	133	(c)	167	(c)	201	(a)
100	(a)	134	(b)	168	(a)	202	(a)
101	(a)	135	(a)	169	(c)	203	(c)
102	(b)	136	(b)	170	(b)	204	(b)
103	(d)	137	(c)	171	(b)	205	(d)
104	(b)	138	(a)	172	(b)	206	(b)
105	(a)	139	(c)	173	(b)	207	(b)
106	(c)	140	(b)	174	(a)	208	(b)
107	(d)	141	(b)	175	(b)	209	(a)
108	(a)	142	(b)	176	(d)	210	(b)
109	(b)	143	(a)	177	(a)	211	(b)
110	(b)	144	(b)	178	(d)	212	(b)
111	(c)	145	(c)	179	(b)	213	(d)
112	(a)	146	(d)	180	(b)	214	(a)
113	(b)	147	(b)	181	(c)	215	(d)
114	(a)	148	(a)	182	(b)	216	(b)

217	(c)	242	(a)	267	(c)	292	(a)
218	(b)	243	(d)	268	(a)	293	(c)
219	(b)	244	(c)	269	(b)	294	(a)
220	(d)	245	(a)	270	(b)	295	(c)
221	(d)	246	(b)	271	(a)	296	(b)
222	(b)	247	(d)	272	(a)	297	(b)
223	(d)	248	(a)	273	(c)	298	(d)
224	(a)	249	(a)	274	(b)	299	(c)
225	(d)	250	(d)	275	(b)	300	(a)
226	(b)	251	(b)	276	(b)	301	(c)
227	(c)	252	(a)	277	(a)	302	(c)
228	(c)	253	(a)	278	(a)	303	(b)
229	(b)	254	(d)	279	(a)	304	(c)
230	(d)	255	(d)	280	(b)	305	(d)
231	(d)	256	(b)	281	(a)	306	(b)
232	(a)	257	(c)	282	(b)	307	(d)
233	(b)	258	(a)	283	(b)	308	(c)
234	(d)	259	(c)	284	(a)	309	(c)
235	(d)	260	(b)	285	(c)	310	(b)
236	(a)	261	(d)	286	(d)	311	(c)
237	(b)	262	(a)	287	(b)	312	(a)
238	(c)	263	(a)	288	(d)	313	(c)
239	(b)	264	(a)	289	(b)	314	(c)
240	(d)	265	(b)	290	(c)		
241	(b)	266	(d)	291	(b)		

Fill in the Blanks

1. Energids or cleavage cells
2. Post embryonic development
3. Larvae
4. Pupa
5. Imago
6. Moulting

7. **Stadium**
8. **Instar**
9. **Coarctate**
10. **Foregut, midgut and hindgut**
11. **Mesenteron**
12. **Trophallaxis**
13. **Flight and Chitin synthesis**
14. **Enteric caecae or Hepatic caecae or Gastric caecae**
15. **Excretion**
16. **Sexual dimorphism**
17. **Bursa copulatrix**

Match the following

1. **a - i, b - iii, c - ii, d - iv**
2. **a - i, b - iv, c - iii, d - ii**
3. **a - ii, b - i, c - iv, d - iii**
4. **a - i, b - iii, c - ii, d - iv**
5. **a - i, b - ii, c - iii, d - iv**
6. **a - i, b - ii, c - iii, d - iv**
7. **a - i, b - ii, c - iii, d - iv**
8. **a - i, b - ii, c - iii, d - iv**
9. **a - iii, b - ii, c - iv, d - i**
10. **a - i, b - iv, c - ii, d - iii**
11. **a - iii, b - i, c - ii, d - iv**

Chapter 4

Ecology

Multiple Choice Questions

1. **Which of the following acts as "nature's scavengers"?**
   (*a*) Insects (*b*) Microorganisms
   (*c*) Man (*d*) Animals

2. **In desert grasslands, which type of animals are relatively more abundant**
   (*a*) Aquatic (*b*) Fossorial
   (*c*) Diurnal (*d*) Arboreal

3. **The supersonic jets cause pollution by the thinning of**
   (*a*) O_2 layer (*b*) O_3 layer
   (*c*) CO_2 layer (*d*) SO_2 layer

4. **Which one of the following organisms is used as indicator of water quality?**
   (*a*) Azospirilluni (*b*) Escherichia
   (*c*) Biggiata (*d*) Chlorella

5. **If there was no CO_2 in the earth's atmosphere, the temperature of earth's surface would be**
   (*a*) Higher than the present
   (*b*) Dependent on the amount of oxygen in the atmosphere
   (*c*) Same as present
   (*d*) Less than the present

6. **Plants such as prosopis, acacia and capparis represent examples of tropical**
   (*a*) Deciduous forests (*b*) Evergreen forests
   (*c*) Grass lands (*d*) Thorn forests

7. **Carbon mono-oxide is a pollutant because**
   (a) Reacts with haemoglobin
   (b) Makes nervous system inactive
   (c) It reacts with O_2
   (d) It inhibits glycolysis

8. **How carbon monoxide, emitted by automobiles, prevents transport of oxygen in the body tissues**
   (a) By forming a stable compound with haemoglobin
   (b) By obstructing the reaction of oxygen with haemoglobin
   (c) By changing oxygen into carbon dioxide
   (d) By destroying the haemoglobin

9. **In a terrestrial ecosystem such as forest, maximum energy is in which trophic level**
   (a) T3 (b) T4
   (c) T1 (d) T2

10. **Which of the following communities is more vulnerable to invasion by outside animals and plants?**
    (a) Temperate forests (b) Oceanic island communities
    (c) Mangroves (d) Tropical evergreen forests

11. **The rate at which light energy is converted into chemical energy of organic molecules is the ecosystems**
    (a) Net secondary productivity (b) Gross primary productivity
    (c) Net primary productivity (d) Gross secondary productivity

12. **The minamata disease in Japan was caused through the pollution of water by**
    (a) Cyanide (b) Methyl isocyanate
    (c) Lead (d) Mercury

13. **Which of the following organism is likely to have more concentration of DDT in its body?**
    (a) Top carnivores (b) Primary producers
    (c) Herbivores (d) Carnivores

14. **Which of the following is pollution related disorder?**
    (a) Silicosis (b) Pneumonicosis
    (c) Fluorosis (d) Leprosis

15. **Energy transfer from one trophic level to other, in a food chain, is**
    (a) 10 per cent (b) 20 per cent
    (c) 1 per cent (d) 2 per cent

16. In 1984, Bhopal gas tragedy was caused due to leakage of

(*a*) Potassium isocyanate
(*b*) Methyl isocyanate
(*c*) Sodium monoxide
(*d*) None of these

17. Which of the following is a secondary pollutant?

(*a*) Pan
(*b*) Aerosol
(*c*) CO
(*d*) CO_2

18. Which part of the world has a high density of organisms?

(*a*) Deciduous forests
(*b*) Tropical rain forests
(*c*) Grasslands
(*d*) Savannahs

19. The maximum biomagnifications would be in which of the following in case of aquatic ecosystem

(*a*) Zooplanktons
(*b*) Phytoplanktons
(*c*) Fishes
(*d*) Birds

20. Which of the following is the use of lichens in case of pollution?

(*a*) They promote pollution
(*b*) Lichens are not related with pollution
(*c*) They treat the polluted water
(*d*) They act as bioindicators of pollutions

21. Land mass occupied by forest is about

(*a*) 30 per cent
(*b*) 60 per cent
(*c*) 11 per cent
(*d*) 22 per cent

22. Which is the result of damage to relative biological effectiveness?

(*a*) High temperature
(*b*) Pollution
(*c*) Radiation
(*d*) Low temperature

23. Which is the reason for highest biomass in aquatic ecosystem?

(*a*) Nano plankton, blue green algae, green algae
(*b*) Sea grass, and slime moulds
(*c*) Benthoic and brown algae
(*d*) Diatoms

24. What is the best pH of the soil for cultivation of plants?

(*a*) 3.4 – 5.4
(*b*) 6.5 – 7.5
(*c*) 4.5 – 8.5
(*d*) 7.5 – 8.5

25. What is BOD?

(a) The amount of O_2 utilized by organisms in water

(b) The amount of O_2 utilized by micro-organisms for decomposition

(c) The total amount of P_2 present in water

(d) All of the above

26. What is the intensity of sound in normal conversation?

(a) 10-20 db (b) 30-60 db

(c) 70-90 db (d) 120-150 db

27. Plant decomposers are

(a) Monera and fungi (b) Fungi and plants

(c) Protista and animalia (d) Animalia and monera

28. Which of the following is absent in polluted water?

(a) Hydrilla (b) Water hyacinth

(c) Larva of stone fly (d) Blue green algae

29. What is true for individuals of same species?

(a) Live in same niche (b) Live in same habitat

(c) Interbreeding (d) Live in different habitat

30. Maximum green house gas released by which of the following country?

(a) India (b) France

(c) USA (d) Britain

31. All insects must communicate in order to

(a) Find a mate and food (b) Survive the winter

(c) Avoid predation (d) All of these

32. Example of carnivorous insects are

(a) Praying mantis (b) Cockroach

(c) Cricket (d) Fruit fly

33. A specific strain of a species adapted to a particular set of environmental conditions is known as

(a) Biotype (b) Ecotype

(c) Allotype (d) Prototype

34. The physical and biological restraints that prevent a species from realizing its biotic potential

(a) Biotic resistance (b) Environmental resistance

(c) Carrying capacity (d) Biological resistance

35. The specific position of a species within a community including utilization of resources both in qualitative and quantitative terms is referred to as

(*a*) Niche (*b*) Habitat

(*c*) Biome (*d*) Ecotone

36. A population of a pest species that differs from other populations of the species in its ability to attack a particular cultivar

(*a*) Biotype (race) (*b*) Biotypic

(*c*) Biotopic (*d*) Biospecies

37. Acid rains are produced by

(*a*) Excess NO_2 and SO_2 from burning fossil fuels

(*b*) Excess production of NH_3 by industry and coal gas

(*c*) Excess release of carbon monoxide by incomplete combustion

(*d*) Excess formation of CO_2 by combustion and animal respiration

38. Which one is not a pollutant normally?

(*a*) Hydrocarbon (*b*) Carbon dioxide

(*c*) Carbon monoxide (*d*) Sulphur dioxide

39. Upper part of sea/aquatic ecosystem contains

(*a*) Plankton (*b*) Nekton

(*c*) Plankton and nekton (*d*) Benthos

40. A simplified linear depiction of trophic relationships that illustrates the flow of energy from producers to herbivores to carnivores

(*a*) Food chain (*b*) Food web

(*c*) Trophic chain (*d*) None of these

41. A situation in which a pest population, after having been suppressed, rebounds to numbers higher than before suppression occurred

(*a*) Secondary pest outbreak (*b*) Pest resurgence

(*c*) Pest outbreak (*d*) Pest resistance

42. An interaction where one of the organisms is harmed by the associated unaffected organisms is called

(*a*) Commensalism (*b*) Amensalism

(*c*) Phoresy (*d*) Mutualism

43. Competition for light, nutrients and space is most severe between

(*a*) Closely related organism growing in different niches

(*b*) Closely related organisms growing in the same area/niche

(*c*) Distantly related organisms growing in the same habitat

(*d*) Distantly related organisms growing in different niches

44. A mutually beneficial association necessary for survival of both partners is

(a) Mutualism/symbiosis (b) Commensalism

(c) Amensalism (d) Both (a) & (b)

45. What is true of ecosystem

(a) Primary consumers are least dependent upon producers

(b) Primary consumers out-number producers

(c) Producers are more than primary consumers

(d) Secondary consumers are the largest and most powerful

46. The complete dependence of one organism over another is called

(a) Amensalism (b) Mutualism

(c) Proto cooperation (d) Commensalism

47. A symbiotic association in which one individual derives benefit and the other is neither helped nor harmed

(a) Commensalism (b) Communalism

(c) Mutualism (d) Parasitism

48. Speciation without geographic isolation is called as

(a) Asympatric speciation (b) Non sympatric speciation

(c) Sympatric speciation (d) None of these

49. Study of associations of organisms in relation to a particular area or habitat is

(a) Syneecology (b) Autecology

(c) Bioecology (d) None of these

50. In an ecosystem, which one shows one-way passage

(a) Free energy (b) Carbon

(c) Nitrogen (d) Potassium

51. Green house effect is warming due to

(a) Infra-red rays reaching earth

(b) Moisture layer in atmosphere

(c) Increase in temperature due to increase in carbon dioxide concentration of atmosphere

(d) Ozone layer of atmosphere

52. Soil conservation is

(a) Conversion of sterile soil into fertile one (b) Aeration of soil

(c) Erosion of soil (d) Protection against loss

53. The relation between algae and fungi in lichen is

(*a*) Symbiosis (*b*) Parasitism

(*c*) Commenalism (*d*) Protocooperation

54. The response of insects to environmental rhythms of light and darkness is known as

(*a*) Phototropism (*b*) Phototrophic

(*c*) Photoperiodism (*d*) None of these

55. Which of the following is example of cannibalism in insects?

(*a*) *Helicoverpa armigera* (*b*) Praying mantids

(*c*) Earwigs (*d*) All of these

56. Genetic changes in populations caused by random phenomenon rather than by selection

(*a*) Gene flow (*b*) Gene pool

(*c*) Genetic drift (*d*) None of these

57. Major aerosol pollutant in jet plane emission is

(*a*) Sulphur dioxide (*b*) Carbon monoxide

(*c*) Methane (*d*) Fluorocarbon

58. Gas released during Bhopal tragedy was

(*a*) Methyl isocyanate (*b*) Potassium isothiocyanate

(*c*) Sodium isothiocyanate (*d*) Ethyl isothiocyante

59. Deforestation will decrease

(*a*) Soil erosion (*b*) Soil fertility

(*c*) Land slides (*d*) Rainfall

60. Pyramid of numbers in a grassland/tree ecosystem

(*a*) Always inverted (*b*) Always upright

(*c*) Both (*a*) & (*b*) (*d*) Spindle-shaped

61. Domestic waste constitutes

(*a*) Non biodegradable pollution (*b*) Biodegradable pollution

(*c*) Effluents (*d*) Air pollution

62. Any unit that includes all the organisms in a given area interacting with the physical environment is called

(*a*) Ecosystem (*b*) Ecotype

(*c*) Ecological balance (*d*) Ecological pyramid

63. The type of diapause in which every individual in every generation enters diapause is called

(a) Obligate diapause
(b) Facultative diapauses
(c) Saprophytic diapauses
(d) None of these

64. Obligate diapause is usually seen in

(a) Univoltine species
(b) Bivoltine species
(c) Multivoltine species
(d) None of these

65. Acid rain is due to increase in atmospheric concentration of

(a) Ozone and dust
(b) CO_2 and CO
(c) SO_2 and CO
(d) SO_2 and NO_2

66. Renewable source of energy is

(a) Biomass
(b) Coal
(c) Petroleum
(d) Kerosene

67. Homeostasis is

(a) Tendency of biological systems to change with change in environment
(b) Tendency of biological systems to resist change
(c) Disturbance of self regulatory system and natural controls
(d) Biotic materials used in homeopathic medicines

68. Adult diapause is seen in which of the following insect

(a) *Leptinotarsa decemlineata*
(b) *Epilachna corrupta*
(c) *Habrobracon brevicornis*
(d) All of these

69. Pupal diapause is seen in

(a) Rice yellow stem borer
(b) Red hairy caterpillar
(c) Castor hairy caterpillar
(d) All of these

70. Viceroy butterfly which resembles with monarch butterfly is thus also avoided by birds is an example of

(a) Batesian mimicry
(b) Ecological mimicry
(c) Biological mimicry
(d) Mullerian mimicry

71. Many species live in the nests of ants as guests are called as

(a) Inquiline
(b) Symbionts
(c) Parasites
(d) None of these

72. Deep black soil is productive due to high proportion of

(a) Sand and zinc
(b) Gravel and calcium
(c) Clay and humus
(d) Silt and earthworm

73. Soil water available to roots is

(*a*) Surface water
(*b*) Hygroscopic water
(*c*) Gravitational water
(*d*) Capillary water

74. Which one is true?

(*a*) Commensalism when none of the interacting populations affect each other
(*b*) Symbiosis when the interaction is useful to both the populations
(*c*) Symbiosis when neither populations affects each other
(*d*) Commensalism when the interaction is useful to both the populations

75. Food chain in which microorganisms breakdown the food formed by primary producers is

(*a*) Parasitic food chain
(*b*) Detritus food chain
(*c*) Consumer food chain
(*d*) Predator food chain

76. In tiger stripe butterfly complex found in south america many species of different families resemble each other and all are distasteful. This is an example of

(*a*) Batesian mimicry
(*b*) Ecological mimicry
(*c*) Biological mimicry
(*d*) Mullerian mimicry

77. Capture recapture method is a__________method of estimates of pest population.

(*a*) Absolute estimates
(*b*) Relative estimates
(*c*) Population indices
(*d*) None of these

78. Which of the following method is used for estimates of pest population?

(*a*) Catch per unit time
(*b*) Line transect
(*c*) Trapping
(*d*) All of these

79. Pick up the correct food chain

(*a*) Grass - Chamelion - Insect - Bird
(*b*) Grass - Fox - Rabbit - Bird
(*c*) Phytoplankton - Zooplankton - Fish
(*d*) Fallen leaves - Bacteria - Insect larvae

80. Most hazardous metal pollutant of automobile exhausts is

(*a*) Mercury
(*b*) Cadmium
(*c*) Lead
(*d*) Copper

81. Which of the following is an example of density independent factors?

(*a*) Predators
(*b*) Parasites
(*c*) Weather
(*d*) None of these

82. Insects which live in places with neither excess nor deficiency of water are known as

(*a*) Hydrocoles (*b*) Mesocoles
(*c*) Exocoles (*d*) None of these

83. Which one of the following is a reason for the population outbreaks?

(*a*) Absence of natural enemies (*b*) Natality may be high
(*c*) Mortality is reduced (*d*) All of these

84. Fertility of soil is measured by its ability to

(*a*) Retain nutrients (*b*) Hold organic materials
(*c*) Hold water (*d*) Support life

85. Petroleum is a

(*a*) Synthetic product (*b*) Renewable resource
(*c*) Nonrenewable resource (*d*) Inconvenient resource

86. Minerals and metals are

(*a*) Biodegradable resources
(*b*) Renewable
(*c*) Non-renewable
(*d*) Renewable and nonrenewable resources

87. Soil fertility is reduced by

(*a*) Crop rotation (*b*) Nitrogen fixing bacteria
(*c*) Decaying organic matter (*d*) Intensive agriculture

88. The insects which are active during the dawn are called

(*a*) Matinal (*b*) Crepuscular
(*c*) Nocturnal (*d*) Diurnal

89. The insects which are active during the day time are called

(*a*) Matinal (*b*) Crepuscular
(*c*) Nocturnal (*d*) Diurnal

90. The insects which are active during the night are called

(*a*) Matinal (*b*) Crepuscular
(*c*) Nocturnal (*d*) Diurnal

91. The insects which are active during the dusk are called

(*a*) Matinal (*b*) Crepuscular
(*c*) Nocturnal (*d*) Diurnal

92. Water is a resource that is

(*a*) Nondegradable nonmaintainabie (*b*) Degradable maintainable

(*c*) Renewable (*d*) Non-renewable

93. Which type of association is found in between entomophilous flower and pollinating agent:

(*a*) Mutualism (*b*) Commensalism

(*c*) Coperation (*d*) Co-evolution

94. Two different species cannot live for long duration in the same niche or habitat this law is

(*a*) Allen's law (*b*) Gause's hypothesis

(*c*) Dollo's rule (*d*) Weisman's theory

95. Bamboo plant is growing in a fir forest then what will be the trophic level of it

(*a*) First trophic level (T1) (*b*) Second trophic level (T2)

(*c*) Third trophic level (T3) (*d*) Fourth trophic level (T4)

96. In aerial net for insect collection the proportion of the diameter of the hoop and the depth of bag is

(*a*) 1 : 3 (*b*) 2 : 5

(*c*) 2 : 7 (*d*) 1 : 2

97. Fluoride pollution mainly affects

(*a*) Brain (*b*) Heart

(*c*) Teeth (*d*) Kidney

98. Mycorrhiza is an example of

(*a*) Symbiotic relationship (*b*) Ectoparasitism

(*c*) Endoparasitism (*d*) Decomposers

99. Certain characteristic demographic developing countries are

(*a*) High fertility, low or rapidly fall rate, rapid population growth and age distribution

(*b*) High fertility, high density, mortality rate and a very young age

(*c*) High infant mortality, low fertility population growth and a very distribution

(*d*) High mortality, high density, uneven growth and a very old age distribution

100. Which of the following is expected to have the highest value ($g/m^2/yr$) in a grassland ecosystem?

(*a*) Secondary production (*b*) Tertiary production

(*c*) Gross production (gp) (*d*) Net production (np)

101. In 1984, the bhopal gas tragedy took place because methyl isocyanate

(a) Reacted with ddt
(b) Reacted with ammonia
(c) Reacted with CO_2
(d) Reacted with water

102. Lead concentration in blood is considered alarming if it is

(a) 20 g/100 ml
(b) 30 g/100 ml
(c) 4 – 6 g/100 ml
(d) 10 g/100 ml

103. In which one of the following pairs is the specific characteristic of a soil not correctly method

(a) Laterite – contains aluminium compound
(b) Terra rosa – most suitable for roses
(c) Chernozems – richest soil in the world
(d) Black soil – rich in calcium carbonate

104. An ecosystem which can be easily damaged but can recover after some time if damaging effect stops will be having

(a) Low stability and high resilience
(b) High stability and low resilience
(c) Low stability and low resilience
(d) High stability and high resilience

105. The maximum growth rate occurs in

(a) Stationary phase
(b) Senescent phase
(c) Lag phase
(d) Exponential phase

106. In which one of the following habitats does the diurnal temperature of soil surface vary most?

(a) Shrub land
(b) Forest
(c) Desert
(d) Grassland

107. In your opinion, which is the most effective way to conserve the plant diversity of an area

(a) By tissue culture method
(b) By creating biosphere reserve
(c) By creating botanical garden
(d) By developing seed bank

108. Prolonged liberal irrigation of agricultural fields is likely to create the problem of

(a) Acidity
(b) Aridity
(c) Salinity
(d) Metal toxicity

109. Which one of the following is not used for disinfection of drinking water?

(a) Chlorine
(b) Ozone
(c) Chloramine
(d) Phenyl

110. Which of the following is not true for a species?

(*a*) Members of a species can interbreed

(*b*) Gene flow does not occur between the populations of a species

(*c*) Each species is reproductively isolated from every other species

(*d*) Variations occur among members of a species

111. Identify the correctly matched pair:

(*a*) Basal convention – biodiversity conservation

(*b*) Kyoto protocol – climatic change

(*c*) Montreal protocol – global warming

(*d*) Ramsar convention – ground water pollution

112. More than 70 per cent of worlds freshwater is contained in

(*a*) Polar ice (*b*) Glaciers and mountains

(*c*) Antartica (*d*) Greenland

113. At which latitude, heat gain through insulation approximately equals heat loss through terrestrial radiation?

(*a*) 22 1/2 north and south (*b*) 40° north and south

(*c*) 42 1/2° north and south (*d*) 66° north and south

114. Animals have the innate ability to escape from predation, examples for the same are given below, select the incorrect example

(*a*) Colour change in chamaeleon

(*b*) Enlargement of body size by swallowing air in puffer fish

(*c*) Poison fangs in snakes

(*d*) Melanism in moths

115. Which one of the following pairs is mismatched?

(*a*) Fossil fuel burning – release of CO_2

(*b*) Nuclear power – radioactive wastes

(*c*) Solar energy – greenhouse effect

(*d*) Biomass burning – release of CO_2

116. In insect collection and preservation, Carbon tetra chloride (Carbona) used as a__________agent.

(*a*) Sticking (*b*) Dehydrating

(*c*) Killing (*d*) Fixing

117. Biodiversity act of India was passed by the parliament in the year

(*a*) 1992 (*b*) 1996

(*c*) 2000 (*d*) 2002

118. Which of the following is considered a hot-spot of biodiversity in India?

(a) Aravalli hills (b) Western ghats

(c) Indo-gangetic plain (d) Eastern ghats

119. Limit of BOD prescribed by Central Pollution Control Board for the discharge of industrial and municipal waste waters into natural surface waters, is

(a) < 30 ppm (b) < 30 ppm

(c) < 10 ppm (d) < 100 ppm

120. Which one of the following is not used for construction of ecological pyramids?

(a) Fresh weight (b) Dry weight

(c) Number of individuals (d) Rate of energy flow

121. Niche overlap indicates

(a) Mutualism between two species

(b) Active cooperation between two species

(c) Two different parasites on the same host

(d) Sharing of one or more resources between the two species (2006)

122. Photochemical smog pollution does not contain

(a) Pan (peroxy acyl nitrate) (b) Ozone

(c) Nitrogen dioxide (d) Carbon dioxide

123. In which one of the following the BOD (Biochemical Oxygen Demand) of Sewage (S), Distillery Effluent (DE), Paper Mill Effluent (PE) And Sugar Mill Effluent (SE) have been arranged in ascending order

(a) SE < PE < S < DE (b) PE < S < SE < DE

(c) S < DE < PE < SE (d) SE < S < PE < DE

124. Geometric representation of age structure is a characteristic of

(a) Population (b) Landscape

(c) Ecosystem (d) Biotic community

125. River water deposits

(a) Loamy soil (b) Alluvial soil

(c) Laterite soil (d) Sandy soil

126. A fertile agricultural soil appears deep coloured at the surface as compared to soil one metre down, the reason for colour of top soil is

(a) More moisture (b) Rich in organic matter

(c) Rich in iron, calcium and magnesium (d) Recent formation

127. Soil particles determine its

(*a*) Texture
(*b*) Field capacity
(*c*) Water holding capacity
(*d*) Soil flora

128. Ultraviolet radiations from sunlight causes a reaction that produces

(*a*) Fluorides
(*b*) Carbon monoxide
(*c*) Sulphur dioxide
(*d*) Ozone

129. American water plant that has become a troublesome water weed in India

(*a*) *Cyperus rotundus*
(*b*) *Eichhornia crassipes*
(*c*) *Typha latifolia*
(*d*) *Trapa bispinosa*

130. Study of inter-relationships between organisms and their environment is

(*a*) Ecology
(*b*) Ecosystem
(*c*) Phytogeography
(*d*) Ethology

131. The sum total of the populations of the same kind of organisms constitute

(*a*) Colony
(*b*) Genus
(*c*) Community
(*d*) Species

132. Soil best suited for plant growth is

(*a*) Clay
(*b*) Loam
(*c*) Sandy
(*d*) Gravel

133. Association of animals when both partners are benefitted

(*a*) Colony
(*b*) Mutualism
(*c*) Commensalism
(*d*) Ammensalism

134. Pyramid of numbers deals with number of

(*a*) Species in an area
(*b*) Individuals in a community
(*c*) Individuals in a tropic-level
(*d*) Subspecies in a community

135. Pyramid of numbers in a pond ecosystem is

(*a*) Irregular
(*b*) Inverted
(*c*) Upright
(*d*) Spindle shaped

136. Which among the following is likely to have the highest levels of DDT depositions in its body?

(*a*) Sea gull
(*b*) Phytoplankton
(*c*) Eel
(*d*) Crab

137. Which one of the following gases contributes maximum to the green house effect' on the earth?

(*a*) Carbon dioxide
(*b*) Chlorofluoro carbon
(*c*) Freon
(*d*) Methane

138. The dominant second trophic level, in a lake ecosystem, is

(a) Phytoplankton
(b) Zooplankton
(c) Benthos
(d) Plankton

139. The major components required for the study of ecosystem are

(a) Host and predators
(b) Temperature and humidity
(c) Biotic and abiotic factor
(d) None of these

140. Stability of ecosystem is maintained by

(a) Diversity of species
(b) Food chain
(c) Food web
(d) Trophic levels

141. When insects remain in dormant stage due to higher temperature than the optimum is called?

(a) Hibernation
(b) Quiscence
(c) Aestivation
(d) None of these

142. For preservation of Cockroaches the pinning region is

(a) Scutellum
(b) Pronotum
(c) Metanotum
(d) Elytra

143. Which one of the following is not a bioindicator of water pollution?

(a) Blood worms
(b) Stone flies
(c) Sewage fungus
(d) Sludge-worms

144. Which one of the following pairs of organisms are exotic species introduced in India?

(a) *Lantana camara* and water hyacinth
(b) Water hyacinth and *Prosopis cinereria*
(c) Nile perch and *Ficus religiosa*
(d) *Ficus religiosa* and *Lantana camara*

145. Which one of the following ecosystem types has the highest annual net primary productivity?

(a) Tropical deciduous forest
(b) Temperate evergreen forest
(c) Temperate deciduous forest
(d) Tropical rain forest

146. The type of vegetation in climate having temperature range 5-20°C and rainfall 150-300 cm will be

(a) Temperate deciduous
(b) Temperate evergreen
(c) Taiga
(d) Grasslands

147. The concept that no two species with identical ecological requirements can coexist in the same place, was proposed by

(*a*) HG Andrewartha (*b*) D Pimentel

(*c*) GF Gause (*d*) LC Birch

148. All the living organisms on earth interacting with the physical environment as a whole are referred to as the

(*a*) Biosphere (*b*) Biome

(*c*) Ecosystem (*d*) Community

149. Among the following which is the best indicator of water pollution due to mixing of human faeces?

(*a*) *Paramecium* (*b*) *Bacillus*

(*c*) *Trypanasoma* (*d*) *E. coli*

150. Among the following which is not an characteristic feature of r-selected species

(*a*) Maturity at later stage (*b*) Large clutch size

(*c*) Small size of offspring (*d*) Maturity at early stage of life

151. Which sampling strategy is best suited for insects that live in the soil?

(*a*) Sex pheromone trap (*b*) Barlese funnel

(*c*) Sweeping (*d*) Light trap

152. Which of these is a primary ecological event?

(*a*) Competition (*b*) Immigration

(*c*) Parasitism (*d*) None of these

153. It has been observed the bacteria secretes toxin at high population density to check the population size. This phenomenon is termed as

(*a*) Quorum sensing (*b*) Population control

(*c*) Alleopathy (*d*) Interspecific competion

154. The accumulation of ddt has decreased the population of pelican ducks because DDT

(*a*) Killed all ducken

(*b*) Decreased availability of ca for egg shells which leads into thin shelled fragile eggs

(*c*) Stopped synthesis of eggs

(*d*) Interfered ca metabolism

155. Hot spots are primarily designated on basis

(*a*) Endemism (*b*) Species diversity

(*c*) Area covered species (*d*) Flowering plant species

156. Phytoplankton spends very little energy on developing protective structures against predator, this suggests that why?

(*a*) Food chain is small

(*b*) There is less competition

(*c*) Productivity of aquatic ecosystem is low

(*d*) Assimilation efficiency is high in aquatic ecosystem

157. Stenohaline organisms can

(*a*) Tolerate narrow range of salt concentration

(*b*) Tolerate wide range of salt

(*c*) Escapes salty conditions

(*d*) Can secrete extra salt to keep salt amount constant

158. Effective population size for completely monogamous species having 40 males and 10 females would be

(*a*) 42 (*b*) 32

(*c*) 20 (*d*) 10

159. Characteristic feature of k-selected species is

(*a*) High intrinsic rate of growth

(*b*) Small size and large number of offsprings

(*c*) Large age at first reproduction

(*d*) Short life span

160. Biodiversity hot spots are characterized on the basis of

(*a*) Endemic flowering plants and threat perception

(*b*) Endemic flowering plants

(*c*) Species of flowering plants

(*d*) Treat perception

161. Correct order for increasing order of primary productivity is

(*a*) Ocean-desert-tropical deciduous forest-tropical rain forest

(*b*) Desert- tropical deciduous- ocean -tropical rain forest

(*c*) Tropical rain forest- ocean- tropical deciduous-deserts

(*d*) Tropical deciduous-desert-ocean-tropical rain forest

162. Type iii survivorship curve is shown by

(*a*) Birds (*b*) Pelagic fishes

(*c*) Humans (*d*) Phytophagous insect

163. The rate of death and expectation of life at various time intervals during the life span of an insect are indicated by

(*a*) Density (*b*) Natality

(*c*) Life table (*d*) Survivorship curve

164. A compilation of bar graphs of the ration of one age group to another in a population is called

(*a*) Age distribution (*b*) Age pyramid

(*c*) Carrying capacity (*d*) Innate capacity

165. A commensalistic relationship in which one kind of animal attaches to another and thereby gains a mode of transportation, is called

(*a*) Philopatry (*b*) Phoresy

(*c*) Amensalism (*d*) Migration

166. The increase in concentration of a substance (usually a pesticide) in animal tissue as related to the animal's higher position in the food chain

(*a*) Bioaccumulation (*b*) Biointensification

(*c*) Biolignification (*d*) Biomagnification

167. In a pyramid of numbers, in a grassland ecosystem, the largest population is that of

(*a*) Producers (*b*) Tertiary consumers

(*c*) Secondary consumers (*d*) Herbivores

168. Which of the following group of gases cause photochemical smog?

(*a*) Ozone, pan and CO (*b*) HC, NO and pan

(*c*) O_2, pan, and NO_2 (*d*) O_3, pan and NO_2

169. Which of the following statements regarding species interdependence are true?

(a) An association of two species where one is benefitted and other remains unaffected called mutualism

(b) An interspecific association where both partners derive benefit from each other is called commensalism

(c) A direct food relation between two species of animals in which one animals kills and feed on another is referred as predation

(d) A relationship between two species of organism where both the partners are benefitted from each other is called symbiosis

(*a*) a and b only (*b*) c and d only

(*c*) a and c only (*d*) b and c only

170. The species of plants that play a vital role in controlling the relative abundance of other species in a community are called

(a) Link species
(b) Pioneer species
(c) Successional species
(d) Keystone species

171. Habitat loss and fragmentation, over exploitation, alien species invasion and co extinction are causes for

(a) Population exploitation
(b) Pollution
(c) Ecological succession
(d) Biodiversity loss

172. The formula of growth rate for population in given time is

(a) Dt/dn=rn
(b) Dt/rn=dn
(c) Dt/rn=dt
(d) Dn/dt=rn

173. One greenhouse gas contributes 14 per cent to total global warming and another contributes 6 per cent These are respectively identified as

(a) N_2O and CO_2
(b) CFCs and N_2O
(c) Methane and CO_2
(d) Methane and CFCs

174. Which of the following is false?

(a) Quantity of biomass in a trophic level at a particular period is called as standing crop
(b) The energy content in a trophic level is determined is determined by considering a few individuals of a species in that trophic level
(c) The succession that occurs in newly cooled lava is called primary succession
(d) Rate of succession is faster in secondary succession

175. Many freshwater animals cannot live for long in sea water and vice versa mainly because of the

(a) Change in n levels
(b) Change in the levels of thermal tolerance
(c) Variations in light intensity
(d) Osmotic problems

176. Which of the following regarding ecological pyramids is not correct?

(a) In most ecosystems, the pyramid of numbers and biomass are up right
(b) In tree dominated ecosystem the pyramid of numbers is inverted
(c) The pyramid of energy expresses mainly the rate of food production
(d) In deep water ecosystem, the pyramid of biomass is upright

177. A high density of elephant population in an area can result in

(a) Intra specific competition
(b) Inter specific competition
(c) Predation on one another
(d) Mutualism

178. For preservation of Beetles the pinning region is

(*a*) Scutellum
(*b*) Pronotum
(*c*) Metanotum
(*d*) Right Elytra

179. Which one of the following statements is correct?

(*a*) Both azotobacter and rhizobium fix atmospheric nitrogen in root nodules of plants
(*b*) Cyanobacteria such as anabaena and nostoc are important mobilizers of phosphates and for plant nutrition in soil
(*c*) At present it is not possible to grow maize without chemical fertilizers
(*d*) Extensive use of chemical fertilizers may lead to eutrophication of nearby water bodies

180. The population of an insect species shows an explosive increase in numbers during rainy season followed by its disappearance at the end of the season. What does this show?

(*a*) The food plants mature and die at the end of the rainy season
(*b*) Its population growth curve is of j-type
(*c*) The population of its predators increases enormously
(*d*) S-shaped or sigmoid growth of this insect

181. A species that interferes with human activities, property or health or is objectionable is known as

(*a*) Pest
(*b*) Host
(*c*) Parasite
(*d*) Parasitoid

182. When insects remain in dormant stage due to lower temperature than the optimum is called

(*a*) Hibernation
(*b*) Quiescence
(*c*) Aestivation
(*d*) None of these

183. For preservation of Bugs the pinning region is

(*a*) Scutellum
(*b*) Pronotum
(*c*) Metanotum
(*d*) Elytra

184. Sounds above what level are considered hazardous noise pollution

(*a*) Above 80 db
(*b*) Above 150 db
(*c*) Above 30 db
(*d*) Above 120 db

185. A disease caused by eating fish contaminated by industrial waste, containing mercury compounds, is called

(*a*) Osteosclerosis
(*b*) Hashimoto's oxidase
(*c*) Bright's disease
(*d*) Minamata disease

186. When huge amount of sewage is dumped into a river, its BOD will

(*a*) Slightly decrease
(*b*) Remain unchanged
(*c*) Increase
(*d*) Decrease

187. If we completely remove the decomposers from an ecosystem, its functioning will be adversely affected, because

(*a*) Mineral movement will be blocked
(*b*) The rate of decomposition will be very high
(*c*) Energy flow will be blocked
(*d*) Herbivores will not receive solar energy

188. In a biotic community, the primary consumers are

(*a*) Detritivores
(*b*) Herbivores
(*c*) Carnivores
(*d*) Omnivores

189. The abundance of a species population, within its habitat, is called

(*a*) Relative density
(*b*) Regional density
(*c*) Absolute density
(*d*) Niche density

190. The Taj Mahal is threatened due to the effect of

(*a*) Oxygen
(*b*) Hydrogen
(*c*) Chlorine
(*d*) Sulphur-dioxide

191. In Minamata Bay Japan the animals which remained free from minamata disease, are

(*a*) Dogs
(*b*) Cats
(*c*) Pigs
(*d*) Rabbits

192. Which of the following pairs is a sedimentary type of biogeochemical cycle?

(*a*) Phosphorus and nitrogen
(*b*) Phosphorus and sulphur
(*c*) Oxygen and nitrogen
(*d*) Phosphorus and carbon dioxide

193. The toxic effect of carbon monoxide is due to its greater affinity for haemoglobin as compared to oxygen approximately by

(*a*) 200 times
(*b*) 1000 times
(*c*) 2 times
(*d*) 20 times

194. Which of the following isotopes is most dangerous to *Homo sapiens*?

(*a*) Phosphorus-32
(*b*) Strontium-90
(*c*) Caesium-137
(*d*) Iodine-131

195. Which of the following is the most stable ecosystem?

(*a*) Mountain (*b*) Ocean

(*c*) Forest (*d*) Desert

196. The primary succession refers to the development of communities on a

(*a*) Forest clearing after devastating fire

(*b*) Newly-exposed habitat with no record of earlier vegetation

(*c*) Freshly cleared crop field

(*d*) Pond, freshly filled with water after a dry phase

197. A dental disease characterized by mottling of teeth is due to the presence of certain chemical element in drinking water Which of the following is that element

(*a*) Fluorine (*b*) Boron

(*c*) Mercury (*d*) Chlorine

198. Which of the following is the main factor of desertification?

(*a*) Over-grazing (*b*) Tourism

(*c*) Irrigated agriculture (*d*) All of these

199. Which one of the following pairs is correctly matched?

(*a*) Parasitism – intra-specific relationship

(*b*) Uricotelism – aquatic habitat

(*c*) Excessive perspiration – xeric adaptation

(*d*) Stream lined body – aquatic adaptation

200. The closely related morphologically similar sympatric populations, but reproductively isolated, are designated as

(*a*) Clones (*b*) Sibling species

(*c*) Clines (*d*) Denies

201. The true statement about 'Green House Effect' is that it is

(*a*) Caused by combination of many gases

(*b*) Caused only by CO_2

(*c*) Caused by CO_2, CFC, CH_4 and NO_2 gases

(*d*) None of these

202. Benthoic animals are those, which

(*a*) Are submerged in area (*b*) Float on the sea surface

(*c*) Are deep dweller in sea (*d*) Are floating (free) organisms

203. Which country has the greatest contribution for the whole formation in ozone layer?

(*a*) Russia (*b*) Japan

(*c*) USA (*d*) Germany

204. The 'niche' of a species is meant for

(*a*) Habitat and specific functions of a species

(*b*) Specific place where an organism lives

(*c*) Specific species function and its competitive power

(*d*) None of these

205. The two great industrial tragedies namely, MIC and Chernobyl tragedies respectively occurred where and at which time

(*a*) Bhopal 1984, Ukrain 1986 (*b*) Bhopal 1986, Russia 1988

(*c*) Bhopal 1984, Ukrain 1990 (*d*) Bhopal 1984, Ukrain 1988

206. The 10 per cent energy transfer law of food chain was given by

(*a*) Lindemann (*b*) Tansley

(*c*) Stanley (*d*) Weismann

207. In coming years, skin related disorders will be more common due to

(*a*) Water pollution (*b*) Depletion of ozone layer

(*c*) Pollutants in air (*d*) Use of detergents

208. During adverse season, therophytes survive by

(*a*) Rhizomes (*b*) Seeds

(*c*) Bulbs (*d*) Corms

209. MAB stands for

(*a*) Mammals and biosphere

(*b*) Mammals and biology programme

(*c*) Man and biology programme

(*d*) Man and biosphere programme

210. Formation of ozone hole is maximum over

(*a*) Europe (*b*) Africa

(*c*) India (*d*) Antarctica

211. Which of the following ecosystem has the highest gross primary productivity?

(*a*) Mangroves (*b*) Rain forest

(*c*) Grassland (*d*) Coral reef

212. Phosphate pollution is caused by

(*a*) Sewage and phosphate rock

(*b*) Sewage and agricultural fertilizers

(*c*) Phosphate rock only

(*d*) Agricultural fertilizers only

213. A narrow band of temperature at which the normal physiological activities of insect takes place is called as

(*a*) Temperature Preferendum

(*b*) Temperature Viability

(*c*) Temperature Acceptability

(*d*) Minimum Effective Temperature

214. The lowest temperature at which organisms can live indefinitely is called as

(*a*) Temperature Preferendum

(*b*) Temperature Viability

(*c*) Temperature Acceptability

(*d*) Minimum Effective Temperature

215. Response of organisms to seasonal and climatic changes is known as

(*a*) Adaptation (*b*) Mutation

(*c*) Acclimatization (*d*) Metamorphosis

216. When the factors act in such a way that constant proportion of population is destroyed irrespective of the number of individuals it contains, they are called as?

(*a*) Catastrophic factor (*b*) An atrophic factor

(*c*) Facultative factor (*d*) Obligatory factor

217. When the factors act in such a way which involve the destruction of a percentage of population with increase as the number of individuals per unit area increases are called as?

(*a*) Catastrophic factor (*b*) An atrophic factor

(*c*) Facultative factor (*d*) Obligatory factor

218. A physiological state of arrested metabolism, growth and development that occurs at one stage in the life cycle under adverse environmental conditions is called

(*a*) Aestivation (*b*) Diapause

(*c*) Hibernation (*d*) None of these

219. Inactivity induced by unfavorable environmental conditions; activity resumes immediately on return of favorable conditions is known as

(a) Aestivation (b) Diapause

(c) Hibernation (d) Quiescence

220. Species which occupy different geographical ranges are termed as

(a) Allopatric species (b) Sympatric species

(c) Biotype (d) Ecotype

221. The measure of the innate ability of a population to survive and reproduce is called

(a) Carrying capacity (b) Biotic potential

(c) Ecological Backlash (d) Fidelity

222. The maximum population density in a given environment will support for a sustained period is called

(a) Carrying capacity (b) Biotic potential

(c) Ecological Backlash (d) Fidelity

223. The counter responses of pest populations or other biotic factors in the environment that devalue the effectiveness of insect management tactics is called as

(a) Carrying capacity (b) Biotic potential

(c) Ecological Backlash (d) Fidelity

224. The total heat energy required to complete a certain stage of development in the life cycle of a species is constant, this is known as

(a) Temperature Preferendum

(b) Thermal Constant

(c) Temperature Acceptability

(d) Minimum Effective Temperature

225. One degree of mean temperature lasting for one day is called

(a) Thermal Constant (b) Temperature Acceptability

(c) Average Effective Temperature (d) Day-degree

226. An ordered depiction of the interconnections that occur among the organisms in food chains in a given environment is called as

(a) Food network (b) Food web

(c) Trophic chain (d) None of these

227. Functional status of an organism in its community is called as

(a) Niche (b) Habitat

(c) Biome (d) Ecotone

228. The phenomenon in which one organism being transported on the body of an individual of another species is called

(*a*) Commensalism (*b*) Amensalism

(*c*) Phoresy (*d*) Mutualism

229. When a major pest is suppressed by a tactic and a minor pest becomes a major pest then this condition is called as?

(*a*) Resistance (*b*) Resurgence

(*c*) Secondary pest outbreak (*d*) Commensalism

230. A species characterized by high reproductive rate and low survival rate is known as

(*a*) P - strategies (*b*) Q - strategies

(*c*) R - strategies (*d*) S - strategies

231. A species characterized by low reproductive rate and high survival rate is known as

(*a*) I - strategies (*b*) J - strategies

(*c*) K - strategies (*d*) L - strategies

232. A subpopulation of interbreeding individuals is called as

(*a*) Species (*b*) Deme

(*c*) Genus (*d*) None of these

233. Groups of species exploiting a common resource in a similar manner are called

(*a*) Genus (*b*) Guilds

(*c*) Family (*d*) Deme

234. The amount of injury that could be produced by an individual pest through its complete life cycle is called as

(*a*) Economic Threshold Level (*b*) Economic Injury Level

(*c*) Insect Equivalent (*d*) Damage Boundary

235. The stability of ecosystem is maintained by

(*a*) Homeostasis (*b*) Diversity of species

(*c*) Food web (*d*) Trophic levels

236. The components required for the study of ecosystem are

(*a*) Soil and water (*b*) Parasites and predators

(*c*) Biotic and Abiotic (*d*) Temperature and humidity

237. The accuracy with which a population estimate follows actual numbers in a population is known as

(a) Phoresy (b) Fidelity
(c) Population dynamics (d) Population density

238. Any energy source that reduces the cost of internal self maintenance of the ecosystem and thereby increases the amount of other energy that can be converted to production is called

(a) Energy consumed (b) Energy subsidy
(c) Energy level (d) Energy loss

239. The data representing the trophic structure at each trophic level by a diagram is

(a) Ecological monogram (b) Ecological graph
(c) Ecological pyramid (d) Ecological shape

240. The opposite sex of the holotype, i.e a paratype (at the time of description) is called the

(a) Acrotype (b) Biotype
(c) Ecotype (d) Allotype

241. A condition when the palatable species imitates an unpleasant one to avoid predation is known as

(a) Batesian mimicry (b) Ecological mimicry
(c) Biological mimicry (d) Mullerian mimicry

242. Organisms living at the bottom of an aquatic ecosystem is called

(a) Nekton (b) Benthos
(c) Plankton (d) All of these

243. Communities which are considered to be transitional between two biomes, are termed as

(a) Niche (b) Habitat
(c) Biome (d) Ecotone

244. The phenomenon of feeding on individuals of same species is called as

(a) Carnivore (b) Herbivore
(c) Polyphagous (d) Cannibalism

245. Maximum number of individuals of an organism that a habitat can maintain at a specific time is called as

(a) Biotic potential (b) Environmental resistance
(c) Carrying capacity (d) Biological capacity

246. A phenomenon when insects become dormant due to unfavorable climatic situation and show no noticeable sign of metabolic action is called as

(*a*) Aestivation (*b*) Hibernation

(*c*) Cryptobiosis (*d*) None of these

247. The daily cycle of dark and light in between day and night and vice versa is termed as

(*a*) Matinal period (*b*) Crepuscular period

(*c*) Circadian rhythm (*d*) All of these

248. Retardation or suspension of morphological growth and development for varying periods to tide over unfavorable conditions of the environment is called

(*a*) Aestivation (*b*) Hibernation

(*c*) Cryptobiosis (*d*) Dormancy

249. Transitional habitat between the freshwater and marine habitats is known as

(*a*) Ecotone (*b*) Estuarine habitat

(*c*) Riverine habitat (*d*) Oceanic habitat

250. Number of individuals added to the population per head per unit of time is called as finite rate of

(*a*) Irregularity (*b*) Increase

(*c*) Decrease (*d*) Constant

251. Maximal rate of enhancement attained by a species at any meticulous amalgamation of temperature, food, moisture etc. is called as innate capacity for

(*a*) Increase in number (*b*) Decrease in number

(*c*) Constant in number (*d*) None of these

252. A systematic tabulation of the survival and causes of mortality that occur in a population throughout a period of time is called as

(*a*) Life chart (*b*) Life table

(*c*) Life graph (*d*) Life diagram

253. Two or more harmful or distasteful species resemble each other. Each of them gets benefit due to similarity because the predator, tasting one, learns to avoid the other also. This type of mimicry is called

(*a*) Batesian mimicry (*b*) Ecological mimicry

(*c*) Biological mimicry (*d*) Mullerian mimicry

254. The cryoprotective compounds which prevent tissue destruction so that insect can survive at freezing temperature are

(a) Glycerol
(b) Sorbitol
(c) Erythriotal
(d) All of these

255. The association of various casts of an insect in a colony is called as

(a) Commensalism
(b) Neutralism
(c) Communalism
(d) None of these

Fill in the Blanks

1. One approach to understanding the number of species and their relative abundance is called__________.

2. The effort to eliminate pests from a geographic area is known as__________.

3. The term 'Ecology' was proposed by__________in 1869.

4. Study of a community in relation to its total environment is known as__________.

5. A group or groups of individuals of anyone kind of organism or species is known as__________.

6. Species with a narrow range of tolerance for temperature is known as__________.

7. Species with a wide range of tolerance for temperature is known as__________.

8. Species with a narrow range of tolerance for food is known as__________.

9. Species with a wide range of tolerance for food is known as__________.

10. The laws governing the number of animals in relation to the area they inhabit is called__________.

11. The number of individuals per unit area or volume is called__________.

12. The theoretical or maximum birth rate under ideal condition is known as__________.

13. A local association of several populations of different species is known as__________.

14. The superficial resemblance of certain animals to other more powerful and more protective ones or like different plant parts for protection is known as__________.

15. Pyramid of numbers in a grassland/tree ecosystem always__________.

16. The term 'niche' was first used by__________.

Match the Following

1. Match the following type of insect with their activity time

(a)	Matinal	i)	Dawn
(b)	Crepuscular	ii)	Night
(c)	Nocturnal	iii)	Dusk
(d)	Diurnal	iv)	Day

2. Match the insects with their pollination type

(a)	Beetles	i)	Myophily
(b)	Ants	ii)	Myrmecophily
(c)	Butterflies	iii)	Palenophily
(d)	Syrphids	iv)	Canthorophily

3. Match the following insects with their stage of diapause

(a)	Grass hopper	i)	Larva
(b)	Pink boll worm	ii)	Egg
(c)	Red hairy caterpillar	iii)	Adult
(d)	*Epilachna* sp.	iv)	Pupa

4. Match the type of diapause with their insects

(a)	Embryo diapause	i)	*Plodia interpunctella*
(b)	Prepupal diapause	ii)	*Culex pipiens*
(c)	Pupal diapause	iii)	*Bombyx mori*
(d)	Imaginal diapause	iv)	*Pieris brassicae*

5. Match the following scientists with term coined by them

(a)	Wheeler	i)	Diapause
(b)	Tansley	ii)	Food chain
(c)	Elton	iii)	Ecosystem
(d)	Grinell	iv)	Niche

Answers

Multiple Choice Questions

1	(b)	21	(a)	41	(b)	61	(a)
2	(b)	22	(b)	42	(b)	62	(a)
3	(b)	23	(c)	43	(b)	63	(a)
4	(b)	24	(b)	44	(a)	64	(a)
5	(d)	25	(b)	45	(c)	65	(d)
6	(d)	26	(b)	46	(b)	66	(a)
7	(a)	27	(a)	47	(a)	67	(b)
8	(a)	28	(c)	48	(c)	68	(d)
9	(c)	29	(c)	49	(a)	69	(d)
10	(d)	30	(c)	50	(a)	70	(a)
11	(b)	31	(d)	51	(c)	71	(a)
12	(d)	32	(a)	52	(d)	72	(c)
13	(a)	33	(b)	53	(a)	73	(d)
14	(c)	34	(b)	54	(c)	74	(b)
15	(a)	35	(a)	55	(d)	75	(b)
16	(b)	36	(a)	56	(c)	76	(d)
17	(a)	37	(a)	57	(d)	77	(a)
18	(b)	38	(b)	58	(a)	78	(d)
19	(c)	39	(a)	59	(d)	79	(c)
20	(d)	40	(a)	60	(b)	80	(c)

81	(c)	115	(c)	149	(d)	183	(a)
82	(b)	116	(c)	150	(a)	184	(d)
83	(d)	117	(d)	151	(b)	185	(d)
84	(d)	118	(b)	152	(a)	186	(c)
85	(c)	119	(a)	153	(a)	187	(a)
86	(d)	120	(a)	154	(b)	188	(b)
87	(d)	121	(d)	155	(a)	189	(d)
88	(a)	122	(d)	156	(b)	190	(d)
89	(d)	123	(b)	157	(a)	191	(d)
90	(c)	124	(a)	158	(b)	192	(b)
91	(b)	125	(b)	159	(c)	193	(a)
92	(c)	126	(b)	160	(a)	194	(b)
93	(d)	127	(a)	161	(a)	195	(b)
94	(b)	128	(d)	162	(d)	196	(b)
95	(a)	129	(b)	163	(c)	197	(a)
96	(d)	130	(a)	164	(b)	198	(a)
97	(c)	131	(d)	165	(b)	199	(d)
98	(a)	132	(b)	166	(d)	200	(c)
99	(a)	133	(b)	167	(a)	201	(c)
100	(c)	134	(c)	168	(d)	202	(c)
101	(d)	135	(c)	169	(b)	203	(c)
102	(b)	136	(a)	170	(d)	204	(a)
103	(d)	137	(a)	171	(d)	205	(a)
104	(a)	138	(b)	172	(d)	206	(a)
105	(d)	139	(c)	173	(b)	207	(b)
106	(c)	140	(a)	174	(b)	208	(a)
107	(b)	141	(c)	175	(d)	209	(d)
108	(c)	142	(b)	176	(a)	210	(d)
109	(d)	143	(b)	177	(a)	211	(b)
110	(b)	144	(a)	178	(d)	212	(b)
111	(b)	145	(d)	179	(d)	213	(a)
112	(a)	146	(a)	180	(b)	214	(d)
113	(b)	147	(c)	181	(a)	215	(c)
114	(c)	148	(a)	182	(a)	216	(a)

217	(c)	227	(a)	237	(b)	247	(b)
218	(b)	228	(c)	238	(b)	248	(d)
219	(d)	229	(c)	239	(c)	249	(b)
220	(a)	230	(c)	240	(d)	250	(b)
221	(b)	231	(c)	241	(a)	251	(a)
222	(a)	232	(b)	242	(b)	252	(b)
223	(c)	233	(b)	243	(d)	253	(d)
224	(b)	234	(c)	244	(d)	254	(d)
225	(d)	235	(b)	245	(c)	255	(a)
226	(b)	236	(c)	246	(c)		

Fill in the Blanks

1. Niche packing
2. Eradication
3. Ernst Haeckel
4. Ecosystem
5. Population
6. Stenothermal
7. Eurythermal
8. Stenophagic
9. Euryphagic
10. Population ecology
11. Population density
12. Biotic potential
13. Community
14. Mimicry
15. Upright
16. Grinell (1917)

Match the following

1. a) - i, b) - iii, c) - ii, d) - iv
2. a) - iv, b) - ii, c) - iii, d) - i
3. a) - ii, b) - i, c) - iv, d) - iii
4. a) - iii, b) - iv, c) - i, d) - ii
5. a) - i, b) - iii, c) - ii, d) - iv

Chapter 5
Biological Control

Multiple Choice Wuestions

1. **Meaning of biological control is**
   (*a*) Control of insect biology
   (*b*) Artificial insect control through biotic factors
   (*c*) Insect control through abiotic factors
   (*d*) All of these

2. **Barnase, Barstar and Bar Genes are used for developing genetically modified**
   (*a*) Brinjal (*b*) Tomato
   (*c*) Cotton (*d*) Mustard

3. **Mycar is a commercial formulation containing**
   (*a*) *Beauveria bassiana* (*b*) *Hirsutella thompsani*
   (*c*) *Verticillium lecani* (*d*) None of these

4. **The method of controlling pest by exposing them to their natural enemy is known as**
   (*a*) Natural control (*b*) Biological control
   (*c*) Integrated Pest control (*d*) Chemical control

5. **An organism which completes its life on one host and kill it after, is known as**
   (*a*) Predator (*b*) Parasite
   (*c*) Parasitoid (*d*) Host

6. **The area under *Bt* cotton in India is about**
   (*a*) 50000 ha (*b*) 100000 ha
   (*c*) 150000 ha (*d*) 200000 ha

7. **Which one of the following is commercial formulation of *Bacillus thuringiensis*?**
   (*a*) Biobit (*b*) Spicturin
   (*c*) Doom (*d*) Dipel

8. **A free living organism kill the host immediately and devour them is known as**
   (*a*) Predator (*b*) Parasite
   (*c*) Parasitiod (*d*) Host

9. **The living organism that serves as food for predator parasite, or pathogen is known as**
   (*a*) Predator (*b*) Parasite
   (*c*) Parasitiod (*d*) Host

10. **In India Genetic Engineering Approval Committee (GEAC) is responsible for monitoring of GM crops comes under ministry of**
    (*a*) Agriculture (*b*) Science and Technology
    (*c*) Fertilizers and Chemicals (*d*) Environment and Forests

11. **Which one of the following is commercial formulation of *Bacillus popilliae*?**
    (*a*) Biobit (*b*) Spicturin
    (*c*) Doom (*d*) Dipel

12. **A form of multiple parasitism in which a parasite preferentially attacks a host that is already parasitized by another species is known as**
    (*a*) Cleptoparasitism (*b*) Ectoparasitoid
    (*c*) Obligate parasitism (*d*) Super parasitism

13. **Parasitoid that lives on the external or outer surface of its host feeding on it and killing it in process of parasitism is known as**
    (*a*) Ectoparasitoid (*b*) Endoparasitoid
    (*c*) Mesoparasitoid (*d*) None of these

14. **The largest area of a transgenic crop in the world is under**
    (*a*) Maize (*b*) Soybean
    (*c*) Cotton (*d*) Rice

15. ***Bacillus popilliae* is mass multiplied on the host**
    (*a*) *Popillia japonica* (*b*) *Spodoptera litura*
    (*c*) *Helicoverpa armigera* (*d*) None of these

16. A parasitoid that lives inside of another organism, feeding on it and killing it in the process of parasitism is known as

(*a*) Ectoparasitoid (*b*) Endoparasitoid

(*c*) Mesoparasitoid (*d*) None of these

17. The species is capable of survival only by parasitism is known as

(*a*) Cleptoparasitism (*b*) Facultative parasites

(*c*) Obligate parasitism (*d*) Super parasitism

18. The largest area under *Bt* cotton in India is in

(*a*) Maharashtra (*b*) Andhra Pradesh

(*c*) Tamil Nadu (*d*) Gujarat

19. Which stage of sugarcane pyrilla is parasitized by *Tetrastichus pyrillae?*

(*a*) Egg (*b*) Larva

(*c*) Pupa (*d*) Adult

20. Importation and release of bio-control agent in an area in which they are not already present, with intent to establish a permanent population is known as__________biological control.

(*a*) Inoculative (*b*) Introducive

(*c*) Inundative (*d*) None of these

21. Release of large numbers of biocontrol agents relative to the number of target species, expection of a rapid effect, there is no implication that the released bio-agents will establish a permanent population is known as__________ biological control.

(*a*) Inoculative (*b*) Introducive

(*c*) Inundative (*d*) None of these

22. Which one of the following is natural enemy of *Lantana camera*

(*a*) *Aspondylla lantana* (*b*) *Ophiomyia lantanae*

(*c*) *Teleonemia scrupulosa* (*d*) All of these

23. Natural enemy of Prickly pear weed, *Opuntia dillenii* is

(*a*) *Aspondylla lantana* (*b*) *Dactylopius opuntiae*

(*c*) *Teleonemia scrupulosa* (*d*) None of these

24. A parasitoid whose nutritional requirements are such that only one can exit in the body of the host is known as__________parasitoid.

(*a*) Solitary (*b*) Gregarious

(*c*) Disperse (*d*) Population

25. The process by which mass-reared agents are released to augment existing population in the field or for reintroduction into habitats where natural enemies have been killed is known as

(*a*) Introduction (*b*) Augmentation

(*c*) Contamination (*d*) Inoculation

26. The share of microbial pesticides in the total world pesticide market is

(*a*) 1 - 2 per cent (*b*) 4 - 6 per cent

(*c*) 8 - 10 per cent (*d*) 10 - 12 per cent

27. Which one of the following is natural enemy of congress grass, *Parthenium hysterophorus*?

(*a*) *Zygogramma bicolorata* (*b*) *Smicronyx lutulentus*

(*c*) *Epiblema strenuana* (*d*) All of these

28. The term "Microbial Control" was first used by

(*a*) Painter, 1951 (*b*) Stern, 1950

(*c*) Steinhans, 1949 (*d*) Bassi, 1835

29. Father of insect pathology is known as

(*a*) Agostino Bassi (*b*) RH Painter

(*c*) KF Stern (*d*) EF Knipling

30. *Bt* cabbage is developed against

(*a*) Diamond back moth (*b*) Cabbage head borer

(*c*) Cabbage butterfly (*d*) Aphid

31. The following is natural enemy of mite

(*a*) *Amblyseius cucumeris* (*b*) *Zygogramma bicolorata*

(*c*) *Crytorhinus lividipennis* (*d*) *Aphelinus mali*

32. Who discovered the muscurdine disease of silk worm by *Beauveria bassiana* in 1835?

(*a*) Agostino Bassi (*b*) RH Painter

(*c*) KF Stern (*d*) EF Knipling

33. Insect parasitoid groups are mainly from order

(*a*) Lepidoptera (*b*) Diptera

(*c*) Coleoptera (*d*) Hymenoptera

34. A strain of *Trichogramma chilonis* has been developed in India which is resistant to

(*a*) Metasystox (*b*) Chlorpyriphos

(*c*) Cartap (*d*) Endosulfan

35. The first record of an insect disease caused by a fungal pathogen is that of

(a) *Nomurea riley* (b) *Metarhizium anisopliae*

(c) *Verticillium lecanii* (d) *Beauveria basiana*

36. Mite species, *Phytosemilus persimilis* is a predator on

(a) *Tetranychus* spp. (b) *Tetrasticus* spp.

(c) *Trichogama* spp. (d) *Bracon* spp.

37. Utilization of disease causing organisms to reduce the population of insect pests below the damaging levels is known as

(a) Microbial control (b) Bacterial control

(c) Natural control (d) Biological control

38. The green lacewing is a predator of

(a) Whitefly (b) White grubs

(c) Aphids (d) Termites

39. Those microorganisms that do not require an insect weakened by external factors to cause infection is known as

(a) Facultative pathogens (b) Obligate pathogens

(c) Potential pathogens (d) None of these

40. The microorganisms may occur outside of the insect and require living insect hosts for survival and replication is known as

(a) Facultative pathogen (b) Obligate pathogen

(c) Potential pathogen (d) None of these

41. Transgenic crops commercially grown first time in India in the year

(a) 2000 (b) 2001

(c) 2002 (d) 2003

42. The stage of the rice gall midge is parasitized by *Platygaster oryzae*

(a) Egg (b) Larva

(c) Pupa (d) Adult

43. The ability of a microorganisms to produce infection is known as

(a) Infectivity (b) Virulence

(c) Pathogenecity (d) Contamination

44. The disease producing power of a microorganisms is known as

(a) Infectivity (b) Virulence

(c) Pathogenecity (d) Contamination

45. Which was the first country to commercialize transgenic crops?

(*a*) USA (*b*) China

(*c*) Canada (*d*) Argentina

46. The beetle, *Zygogramma bicolorata*, introduced for the control of congress grass, is also found to be a pest on

(*a*) Rice (*b*) Sugarcane

(*c*) Cotton (*d*) Sunflower

47. The disease producing power of a group of species of microorganisms is known as

(*a*) Infectivity (*b*) Virulence

(*c*) Pathogenecity (*d*) Contamination

48. *Bacillus thruingienesis* is a__________bacterium.

(*a*) Motile (*b*) Gram positive

(*c*) Spore forming (*d*) All of these

49. One of the *Bt* cotton varieties approved for commercial cultivation under north Indian conditions is

(*a*) MECH 12 (*b*) MECH 162

(*c*) MECH 184 (*d*) MRC 6304

50. Which one of the following is bio-control agent of cotton white fly?

(*a*) *Crytorhinus lividipennis* (*b*) *Encarsia formosa*

(*c*) *Aphelinus mali* (*d*) All of these

51. Milky disease of grab of Japanese beetle, *Papillia japonica* is caused by

(*a*) *Bacillus thruingienesis* (*b*) *Bacillus papillae*

(*c*) *Beauveria basiana* (*d*) None of these

52. The most important toxin produced by *Bt* is

(*a*) α - endotoxin (*b*) β - endotoxin

(*c*) δ - endotoxin (*d*) γ - endotoxin

53. The descending order of the area of cultivation of transgenic crops for the trait specific is

(*a*) Herbicide tolerant > Insect resistant > Herbicide tolerant + Insect resistant > Virus resistant

(*b*) Insect resistant > Herbicide tolerant > Nematode resistant > Virus resistant

(*c*) Herbicide tolerant > Nematode resistant > Insect resistant > Virus resistant

(*d*) None of these

54. The All India Co-ordinated Research Project on Biological Control of Crop Pests and Weeds (AICRPBC) was started in

(a) 1972 (b) 1975
(c) 1977 (d) 1981

55. The gut pH of a lepidopteran larvae is

(a) Acidic (b) Neutral
(c) Alkaline (d) Not detected

56. The first major commercial *Bt* product is

(a) Dipel (b) Dimilin
(c) Mycer (d) Elcar

57. The countries leading in the commercial cultivation of transgenic crops

(a) USA > Argentina > Brazil > Canada
(b) Argentina > Brazil > Canada > USA
(c) USA > China > India > Canada
(d) USA > Brazil > China > India

58. The first successful attempt to utilize biological control involved the importation of

(a) Vedalia beetle (b) *Chrysoperla* spp.
(c) *Trichogramma* spp. (d) Ladybird beetle

59. The first baculoviruses to be developed for commercial use was

(a) Dipel (b) Dimilin
(c) Mycer (d) Elcar

60. The diseases of insects caused by fungi are termed as

(a) Itching (b) Fungoses
(c) Mycoses (d) Ufra

61. The descending order of area under transgenic crops

(a) Soybean > Maize > Cotton > Canola
(b) Soybean > Cotton > Rice > Maize
(c) Cotton > Maize > Soybean > Rice
(d) Soybean > Rice > Cotton > Canola

62. Cassava mealy bug, *Phenococcus manihoti* in Africa was successfully controlled by importation and augmentation of *Epidinocarsis lopezi* from

(a) Paraguay (b) Brazil
(c) Europe (d) Russia

63. The green muscardine fungus is

(*a*) *Beuveria bassiana* (*b*) *Metarhizium anisopliae*

(*c*) *Trichoderma viridi* (*d*) *T. ludeni*

64. The white muscardine fungus is

(*a*) *Beuveria bassiana* (*b*) *Metarhizium anisopliae*

(*c*) *Trichoderma viridi* (*d*) *T. ludeni*

65. Seeds of transgenic plant varieties for research purpose will be imported only through

(*a*) NBPGR

(*b*) DBT

(*c*) DST

(*d*) State Department of Agriculture

66. Cassava mealy bug, *Phenococcus manihoti* (Homoptera : Pseudococcidae) in Africa was successfully controlled by importation and augmentation of

(*a*) *Cryptolaemus montrouzieri* (*b*) *Crytorhinus lividipennis*

(*c*) *Apoanagyrus (Epidinocarsis) lopezi* (*d*) *Aphelinus mali*

67. Entomopatogenic nematodes cause disease to insect are

(*a*) *Steinernema* (*b*) *Heterorhabditties*

(*c*) *Globodera* (*d*) Both (*a*) & (*b*)

68. The first identified Entomopathogenic nematodes is *Steinernema glaseri*, cause disease to

(*a*) *Popillia japonica* (*b*) *Papilio demolius*

(*c*) *Bombyx mori* (*d*) *Coccinella septempunctata*

69. The second generation of *Bt* cotton contained genes of

(*a*) Cry 1Ac and Cry 2Ab (bollgurd 2) (*b*) Cry 1Aa and Cry 2Aa

(*c*) Cry 1Ad and Cry 2Ab (*d*) Cry 1Ac and Cry 2Aa

70. Dr. Hans R. Herren was awarded the world food prize for his work on biological control of cassava mealy bug in the year

(*a*) 1995 (*b*) 1990

(*c*) 1982 (*d*) 1980

71. Protozoa *Frinocystis tribolii* infects

(*a*) *Tribolium castaneum* (*b*) *Sitophilus oryzae*

(*c*) *Rizopertha dominica* (*d*) *Earias vitella*

72. *Epiricania melanolenca* is a nymphal and adult Parasitoid of

(*a*) *Pyrilla perpusilla* (*b*) *Popillia japonica*

(*c*) *Papilio demolius* (*d*) *Bombyx mori*

73. *Bt* cotton hybrids first approved by GEAC for commercial cultivation in 2002 are

(*a*) MECH 12, MECH 162 and MECH 184 (*b*) Ankur 09 and Ankur 651

(*c*) Bunny *Bt* 4 and Mallika *Bt* (*d*) Tulasi 4 and Tulasi 117

74. *Platygaster oryzae* is the bio-control agent of

(*a*) Rice Stem Borer (*b*) Cotton Bollworm

(*c*) Fruit Borer (*d*) Rice Gall Midge

75. *Epiricania melanoleuca* an insect parasitoid belongs to the family Epiricanidae under the order

(*a*) Lepidoptera (*b*) Coleoptera

(*c*) Diptera (*d*) Hymenoptera

76. Most widely used soil bacterium in insect control is

(*a*) *Bacillus thuringiensis* (*b*) *Bacillus papillae*

(*c*) *Bacillus cereus* (*d*) *Bacillus pluton*

77. Chronological steps in the revised guidelines for the research in transgenic plants in India are

(*a*) Applicant - IBSC - RCGM - GEAC - Commercial release

(*b*) Applicant - RCGM - IBSC - GEAC - Commercial release

(*c*) Applicant - GEAC - IBSC - RCGM - Commercial release

(*d*) Applicant - IBSC - GEAC - RCGM - Commercial release

78. Egg parasitoid *Chelonus blackburni* is used to check the population of

(*a*) Potato Tuber Moth (*b*) Cut Worm

(*c*) Aphids (*d*) Jassids

79. *Bacillus thuringiensis* produces

(*a*) Delta Endotoxin (*b*) Alpha Exotoxin

(*c*) Beta Exotoxin (*d*) All of these

80. *Bacillus thuringiensis* was discovered by

(*a*) Berliner (*b*) Dulmage

(*c*) Ishiwata (*d*) Burges

81. The only pest that has developed resistance to *Bacillus thuringiensis* in the field conditions

(a) *Helicoverpa armigera* (b) *Plutella xylostella*
(c) *Spodoptera litura* (d) None of these

82. Which of the following fungus is effective on root grubs?

(a) *Fusarium* sp. (b) *Beauveria* sp.
(c) *Nomuraea rileyi* (d) *Metarrhizium anisopliae*

83. *Bacillus thuringiensis* was first isolated from diseased larvae of

(a) American bollworm (b) Corn borer
(c) Silkworm (d) Diamond Back Moth

84. Ladybird beetle is predator of which of the following insects

(a) Gram pod borer (b) Cottony cushion scale
(c) Semilooper (d) Grasshopper

85. The most dominant trait under commercial cultivation of transgenic crops

(a) Herbicide tolerant (b) Nematode resistant
(c) Insect resistant (d) Virus resistant

86. Which of the following insect is controlled by the Nuclear Polyhedral Virus?

(a) *Helicoverpa armigera* (b) *Spodoptera litura*
(c) Red Hairy Caterpillar (d) All of these

87. Name of the predator generally used by the farmers

(a) *Rodalia* sp. (b) *Trichogramma* sp.
(c) *Crysoperlla* sp. (d) None of these

88. Example of pupal parasitoid

(a) *Elasmus* (b) *Tetrasticus*
(c) *Xanthopimpla* (d) All of these

89. The virus used for the control of top shoot borer

(a) NPV (b) GV
(c) CPV (d) All of these

90. San Jose Scale has been successfully controlled in India by the importation of

(a) *Cyrtorhinus lividipennis* (b) *Encarsia perniciosi*
(c) *Coccinella septempunctata* (d) *Trichogramma chilonis*

91. Which order having maximum insect predators?

(a) Isoptera (b) Hymenoptera
(c) Neuroptera (d) Lepidoptera

92. Which of the following predators does not belong to the order Coleoptera?

(a) Chilomenes (b) Scymnus
(c) Coccinella (d) Syrphus

93. Commercially granulosis virus (GV) used for the control of

(a) *Chilo infuscatellus* (b) *Helicoverpa armigera*
(c) *Spodoptera litura* (d) All of these

94. Which one of the following is a parasitoid of apple wooly aphid, *Eriosoma lanigerum?*

(a) *Cryptolaemus montrouzieri* (b) *Aphelinus mali*
(c) *Spogossia bezziana* (d) None of these

95. Which of the following is a neuropteran predator?

(a) Chrysoperla (b) Cryptolaemus
(c) Menochilus (d) Brumus

96. Predator of American bollworm is

(a) *Chrysoperla* spp. (b) *Trichogramma* spp.
(c) *Nematospora gossypii* (d) None of these

97. Baculoviruses contain

(a) Single stranded DNA (b) Double stranded DNA
(c) Single stranded RNA (d) Single stranded RNA

98. White tailed mealy bug, *Ferrisia virgata* (Cockerell) - is used for the rearing of following coccinellid predator

(a) *Cryptolaemus montrouzieri* (b) *Brumoides suturalis*
(c) *Scymnus nubilus* (d) All of these

99. The recommended dose of NPV for the control of *Helicoverpa armigera* under field condition is

(a) 125 - 250 LE (b) 250 - 300 LE
(c) 250 - 500 LE (d) 750 - 1000 LE

100. Example of *Bt* transgenic cotton

(a) Varatalaxmi (b) Jayadhar
(c) NH - 44 (d) Bollgourd

101. Nuclear Polyhedrosis Virus (NPV) is the most effective for control of

(a) *Chilo partellus* (b) *Pectinophora gossypiella*
(c) *Helicoverpa armigera* (d) *Diacrasia oblique*

102. A parasitic species grows parasitically on other parasitic species, this phenomenon is called as

(*a*) Complex parasitism (*b*) Multiple parasitism

(*c*) Hyper parasitism (*d*) None of these

103. Broad spectrum bacteria used against insect pest

(*a*) *Albugo* spp. (*b*) *Fusarium* spp.

(*c*) *Bacillus thuringiensis* (*d*) *Bacillus popillae*

104. Trade name "Achook" is made from

(*a*) Rose (*b*) Chrysanthemum

(*c*) Neem (*d*) Sabadilla

105. Which one of the following is/are entomopathogenic fungi?

(*a*) *Beauveria bassiana* (*b*) *Metarrhizium anisopliae*

(*c*) *Nomurea rileyi* (*d*) All of these

106. The type of parasitism where the host is attacked by two or more species of parasitoids is called

(*a*) Super parasitism (*b*) Multiple parasitism

(*c*) Simple parasitism (*d*) Hyper parasitism

107. Nicotine is commercially available as

(*a*) Nicotine sulphate (*b*) Nicotine phosphate

(*c*) Nicotine (*d*) Acetylcholine

108. Which part of congress grass is used as pesticide?

(*a*) Root (*b*) Leaf

(*c*) Flower (*d*) Stem

109. *Metarrhizium anisopliae* used against sugarcane leaf hopper and rhinoceros beetle are commercially available under the trade name of

(*a*) Biomax (*b*) Mycar

(*c*) Elcar (*d*) None of these

110. Which of the following is a myco-acaricide?

(*a*) *Hirsutella* spp. (*b*) *Neozygites* spp.

(*c*) *Sporothrix* spp. (*d*) None of these

111. Trade name of *Bt* is

(*a*) Biolep (*b*) Neem

(*c*) Biomax (*d*) Achook

112. 1st *Bt* cotton variety was released in

(*a*) 1994 (*b*) 1996

(*c*) 1997 (*d*) 2000

113. Which of the following is an obligate pathogen?

(*a*) *Bacillus thuringiensis* (*b*) *Bacillus popilliae*

(*c*) *Bacillus cereus* (*d*) All of these

114. A type of parasitism where more individuals of the same species are present in a single host that can complete development in a normal way is known as

(*a*) Super parasitism (*b*) Multiple parasitism

(*c*) Simple parasitism (*d*) Hyper parasitism

115. The first example of Biological control of insect pest was found in which country?

(*a*) China (*b*) India

(*c*) Pakistan (*d*) USA

116. Jowar stem borer biologically controlled by

(*a*) *Encarsia fomusa* (*b*) *Apanteles flavipes*

(*c*) Both (*a*) & (*b*) (*d*) None of these

117. Which of the following is both acaricide and insecticide?

(*a*) *Vertillum lecanii* (*b*) *Bacillus thuringiensis*

(*c*) Both (*a*) & (*b*) (*d*) None of these

118. The important family of Hymenoptera which contain effective parasitoids

(*a*) Braconidae (*b*) Trichogrammatidae

(*c*) Aphelinidae (*d*) All of these

119. Example of fungi predacious on nematodes

(*a*) *Didymella* spp. (*b*) *Dectyaria* spp.

(*c*) *Metarhyzeum* spp. (*d*) None of these

120. Which of the following is an egg-larval parasitoid?

(*a*) *Chelonus blackburni* (*b*) *Chrysoperla* spp.

(*c*) *Trichogramma* spp. (*d*) *Epiricania melanoleuca*

121. Which one of the following is an advantage of microbial pesticides

(*a*) They are species specific (*b*) Less possibility of resistance development

(*c*) Safe and eco-friendly (*d*) All of these

122. Which precaution is necessary for introducing a parasitoid or predator in a locality?

(*a*) It should not feed on plants

(*b*) It should not feed on beneficial insects

(*c*) It should not impair the effectiveness of native parasitoid

(*d*) All of these

123. Inundative release is a part of

(*a*) Host plant resistance (*b*) Chemical control

(*c*) Mechanical control (*d*) Biological control

124. The term Biological Control was coined by

(*a*) Smith (*b*) Geier and Clark

(*c*) Bartlet (*d*) None of these

125. The entomogenous fungi tested under field condition in India

(*a*) *Beauveria bassiana* (*b*) *Entomophthora virulenta*

(*c*) *Metarrhizium anisopliae* (*d*) All of these

126. The first transformation of deltaendotoxin gene from *Bt* was done in Belgium in the year 1983 in

(*a*) Cotton (*b*) Tobacco

(*c*) Potato (*d*) Soybean

127. In baculoviruses tolerance to ultraviolet radiations can be increased by the addition of

(*a*) Activated charcoal (*b*) Egg albumin

(*c*) Ranipal (*d*) All of these

128. Largest producer of pyrethrum in the world is

(*a*) India (*b*) USA

(*c*) Nigeria (*d*) Kenya

129. Black scale insects of citrus is controlled by

(*a*) *Rodolia cardinalis* (*b*) *Metaphycus loundsbury*

(*c*) Both (*a*) & (*b*) (*d*) None of these

130. White fly in greenhouses may be controlled by

(*a*) *Encarsia formosa* (*b*) *Chrysoperla* spp.

(*c*) *Coccinella* spp. (*d*) All of these

131. A typical symptom of virus infection in insects is

(a) The infected larvae climb to higher position and dead larvae usually hangs upside down

(b) The integument becomes fragile

(c) Rupture easily and emit foul smell

(d) All of these

132. Which of the following is obtained from roots?

(a) Sabadilla (b) Neem

(c) Limonene (d) Rotenone

133. For the control of red locust in Mauritius the bird exported from India is

(a) Wild crow (b) Maynah

(c) Vulture (d) House sparrow

134. The major techniques involved in a biological control programme are

(a) Importation and Conservation

(b) Conservation and Augmentation

(c) Importation and Augmentation

(d) All of these

135. Which of the following is a limitation in the use of Baculoviruses?

(a) Lacks quick knock down affect

(b) Inactivated by sun light

(c) Early instars are more susceptible than later instars

(d) All of these

136. Which of the following is a muscle poison?

(a) Neem (b) Ryanodine

(c) Nicotine (d) Pyrthrins

137. The actions to preserve and increase natural enemies by environmental manipulation is known as

(a) Importation (b) Conservation

(c) Augmentation (d) None of these

138. The efforts to increase population of natural enemies either by propagation and release or by environmental manipulation is known as

(a) Importation (b) Conservation

(c) Augmentation (d) None of these

139. ***Rodalia cardinalis*** **was introduced into India in 1929 for the control of**

(*a*) Gram pod borer
(*b*) Cottony cushion scale
(*c*) Semilooper
(*d*) Grasshopper

140. In 1898 ***Cryptolemus*** **sp. was introduced into India from Australia for the control of**

(*a*) Gram pod borer
(*b*) Cottony cushion scale
(*c*) Citrus mealy bug
(*d*) Grasshopper

141. ***Bacillus thuringiensis*** **was first isolated from silkworm in**

(*a*) 1900
(*b*) 1902
(*c*) 1911
(*d*) 1912

142. Which of the following is true regarding ***Bacillus thuringiensis***

(*a*) It is rod shaped gram positive bacteria
(*b*) Spore forming, crystalliferous bacteria
(*c*) Aerobic and facultative bacteria
(*d*) All of these

143. Which of the following is true regarding ***Bt*** **mode of action**

(*a*) It is a stomach poison
(*b*) It require alkaline gut pH
(*c*) It takes 3 - 4 days to get insect killed
(*d*) All of these

144. The transgenic (Bt) vegetable which has been approved for large-scale field trials in India is

(*a*) Bhindi
(*b*) Brinjal
(*c*) Tomato
(*d*) Potato

145. The National Bureau of Agriculturally Important Insects formerly Project Directorate of Biological Control (PDBC) came into being in

(*a*) 1963
(*b*) 1973
(*c*) 1983
(*d*) 1993

146. The National Bureau of Agriculturally Important Insects formerly Project Directorate of Biological Control (PDBC) is located at

(*a*) Hyderabad
(*b*) Bangaluru
(*c*) Mumbai
(*d*) New Delhi

147. In case of NPV, the number of days required from the time of insertion of virus and death of host insect is

(*a*) 3 week
(*b*) 4 - 6 days
(*c*) 1 month
(*d*) 1 day

148. The part of the neem which possess maximum active ingredient of insecticidal activity

(*a*) Leaf (*b*) Roots

(*c*) Kernel (*d*) Flowers

149. *Conobathra aphidivora* is a potential predator of

(*a*) Jasid

(*b*) Whitefly

(*c*) Mustard aphid

(*d*) Sugarcane white woolly aphid

150. UV protectant added in NPV formulations to protect from direct sunlight is

(*a*) Black ink (*b*) Charcole

(*c*) Plant extracts (*d*) Both (*a*) & (*b*)

151. Neem is native to

(*a*) India (*b*) USA

(*c*) China (*d*) Brazil

152. Which of the following is commercial neem based formulation?

(*a*) Achook (*b*) Margosan - O

(*c*) Neemguard (*d*) All of these

153. *Bt* was isolated by Berliner in 1915 from diseased larvae of

(*a*) Silk moth (*b*) Mediterranean flour moth

(*c*) Fruit sucking moth (*d*) Helicopter moth

154. In 1989, Hofte and Whiteley proposed a nomenclature and classification scheme for crystal proteins based on their

(*a*) Host range (*b*) Primary structure

(*c*) Colour (*d*) Both (*a*) & (*b*)

155. The type of release of natural enemies in which control is expected from the progeny and subsequent generations and not from the release itself

(*a*) Inundative release (*b*) Cumulative release

(*c*) Inoculative release (*d*) None of these

156. Which one of the following is an attribute of an effective natural enemy?

(*a*) High searching capacity (*b*) High host specificity

(*c*) Short life cycle and high fecundity (*d*) All of these

157. First *Bt* based major commercial product is

(*a*) Dipel (*b*) Doom

(*c*) Elcar (*d*) Mycar

158. First commercial product based on milky disease bacteria is

(*a*) Dipel (*b*) Doom

(*c*) Elcar (*d*) Mycar

159. Rice meal moth is widely used for the rearing of the following parasitoids

(*a*) *Trichogramma* spp. (*b*) *Crysoperla* spp.

(*c*) *Rodalia* spp. (*d*) All of these

160. Which of the following families contain effective predators of insect pests?

(*a*) Coccinellidae (*b*) Reduvidae

(*c*) Chrysopidae (*d*) All of these

161. First NPV based commercial product is

(*a*) Dipel (*b*) Doom

(*c*) Elcar (*d*) Mycar

162. First ever proprietary myco-acaricide in the world is

(*a*) Dipel (*b*) Doom

(*c*) Elcar (*d*) Mycar

163. Which of the following laboratory host is used for the multiplication of egg parasitoids?

(*a*) Rice weevil (*b*) Rice butterfly

(*c*) Rice meal moth (*d*) Indian meal moth

164. Which one on the following is most threatening to the rice meal moth mass multiplication?

(*a*) *Tribolium* spp. (*b*) *Alphitobius diaperinus*

(*c*) Mites (*d*) None of these

165. Mycar is the first ever proprietary myco-acaricide in the world, recommended for use against

(*a*) Eriophyids on citrus (*b*) Tetranychids on chilli

(*c*) Tersonemids on jute (*d*) Tenuipalpids on mango

166. Piericidins are metabolites produced by

(*a*) *Tetranychus* sp. (*b*) *Streptomyces* sp.

(*c*) *Bacillus* sp. (*d*) *Xanthomonus* sp.

167. DD-136 was first isolated from

(*a*) Silk moth (*b*) Codling moth

(*c*) Fruit sucking moth (*d*) Helicopter moth

168. For mass multiplication of DD-136 nematode the common host used is

(*a*) Silk moth
(*b*) Lesser wax moth
(*c*) Fruit sucking moth
(*d*) Helicopter moth

169. Which of the following is not correct with regard to pyrethrins?

(*a*) They have quick knock down affect
(*b*) They are degraded by sunlight
(*c*) They are nerve poisons
(*d*) They are obtained from roots of *Chrysanthemum* spp.

170. Egg parasitoids belong to the family

(*a*) Trichogrammatidae
(*b*) Scleoniidae
(*c*) Evanidae
(*d*) All of these

171. In the world, first transgenic plant was developed in tobacco against the insect pest

(*a*) Tobacco caterpillar, *Spodoptera litura*
(*b*) Tobacco hornworm, *Manduca sexta*
(*c*) Gram pod borer, *Helicoverpa armigera*
(*d*) White fly, *Bemisia tabaci*

172. The *Bt* cotton developed for resistance to bollworms using an indigenous variety is

(*a*) Taranga
(*b*) Bikaneri Narma
(*c*) Bikram
(*d*) Abadhita

173. The classical example of successful biological agent

(*a*) *Rodalia*
(*b*) *Apanteles*
(*c*) *Bracon*
(*d*) *Trichogramma*

174. To which of the following groups do insect pathogens belong?

(*a*) Virus and Bacteria
(*b*) Bacteria and Fungi
(*c*) Fungi and Protozoa
(*d*) All of these

175. KNOCK W.P. is registered and commercialized formulation of bio-insecticide of

(*a*) Fungus strain
(*b*) *Bt* strain
(*c*) NPV strain
(*d*) None of these

176. Which is the larvicidal fish?

(*a*) *Gambusia* spp.
(*b*) *Barbus* spp.
(*c*) *Labeo* spp.
(*d*) *Catla* spp.

177. NPV is used for the control of

(*a*) Gundhi bug (*b*) Pyrilla

(*c*) Mealy bug (*d*) Gram pod borer

178. *Apanteles* is a parasitoid of

(*a*) Egg (*b*) Larva

(*c*) Pupa (*d*) Adult

179. The ultimate control through biotic agencies is expected after

(*a*) One generation (*b*) Two generations

(*c*) Three generations (*d*) More than three generations

180. The recommended dose of NPV for the control of *Spodoptera litura* under field condition is

(*a*) 125 to 250 LE/ha (*b*) 250 to 500 LE/ha

(*c*) 500 to 700 LE/ha (*d*) 750 to 1000 LE/ha

181. When did the first successful achievement of biological control of weed occurs

(*a*) 1900 (*b*) 1902

(*c*) 1904 (*d*) 1906

182. Most common example of insects which is used as bioagent of weeds belong to the order

(*a*) Lepidoptera (*b*) Coleoptera

(*c*) Hymenoptera (*d*) All of these

183. Standard size of a 'Tricho card' is

(*a*) 10 x 5 cm (*b*) 12 x 7.5 cm

(*c*) 15 x 7.5 cm (*d*) 12 x 6 cm

184. On a standard size 'Tricho card' 1 cc eggs are glued approximately in no. is__________

(*a*) 10000 (*b*) 15000

(*c*) 20000 (*d*) 30000

185. Which of the following factor(s) affect the biological control of weeds?

(*a*) Choice of bio-agents (*b*) Climatic condition

(*c*) Aggressiveness of the weeds (*d*) All of these

186. Which of the following is the biological control of insects?

(*a*) Use of predators (*b*) Use of living organisms

(*c*) Use of parasites (*d*) All of these

187. Dose of *Trichogramma chilonis* for controlling cotton bollworm is__________ parasitized eggs/ha.

(a) 50000 (b) 100000

(c) 150000 (d) 200000

188. The__________stage of green lacewing attacks the host

(a) Egg (b) Larva

(c) Pupa (d) Adult

Fill in the Blanks

1. *Cotesia plutella* is parasitoid belongs to the family__________.
2. *Goniozus nephantidis* is a parasitoid belongs to the family__________.
3. Dose of *T. chilonis* for controlling Rice yellow stem borer is__________parasitized eggs/ha.
4. Predators have a__________host range than parasitoid.
5. *Capidosoma koehleri* is a promising parasitoid of__________.

Match the Following

1. Match the parasitoid with their type

(a) *Trichogramma chilonis*	i) Larval parasitoid
(b) *Goniozus nephantidis*	ii) Egg parasitoid
(c) *Xanthopimpla punctata*	iii) Ecto parasitoid
(d) *Epiricania melanoleuca*	iv) Pupal parasitoid

2. Match the following predators with their order

(a) *Dipha aphidivora*	i) Hemiptera
(b) *Chrysoperla carnea*	ii) Coleoptera
(c) *Rodolia cardinalis*	iii) Neuroptera
(d) *Crytorhinus lividipennis*	iv) Lepidoptera

3. Match the following predators with their targeted pests

(a) *Dipha aphidivora*	i) Rice Brown Plant Hopper
(b) *Chrysoperla carnea*	ii) Sucking pests of vegetables
(c) *Rodolia cardinalis*	iii) Sugarcane woolly aphid
(d) *Crytorhinus lividipennis*	iv) Cottony cushion scale

4. Match the weeds with their biocontrol agents

(a)	*Parthenium hysterophorus*	i)	*Neochetina bruchi*
(b)	*Lantana camera*	ii)	*Ophiomyia lantanae*
(c)	*Opuntia dilleni*	iii)	*Dactylopius opuntiae*
(d)	*Eichornia crassipes*	iv)	*Zygogramma bicolorata*

5. Match the following microorganism with their commercial formulation

(a)	*Hirsutella thompsoni*	i)	Mycar
(b)	*Metarhizium anisopliae*	ii)	Vertilec
(c)	*Verticillium lecanii*	iii)	Biomax
(d)	*Beauveria Basiana*	iv)	Boverin

6. Match the following biocontrol agencies with their host insect

(a)	*Oecophylla smaragdina*	i)	Citrus mealy bug
(b)	*Spogosia bezziana*	ii)	Black headed caterpillar
(c)	*Cryptolaemus montrouzieri*	iii)	Cottony cushion scale
(d)	*Rodolia cardinalis*	iv)	Leaf chewing insects

7. Match the following insects with their rearing feed

(a)	Aphid	i)	Potato
(b)	Mealy bug	ii)	Cowpea seed
(c)	Rice meal moth	iii)	Cotton seed
(d)	Red cotton bug	iv)	Broken grain of Sorghum

Answers

Multiple Choice Questions

1	(b)	21	(c)	41	(a)	61	(a)
2	(d)	22	(b)	42	(b)	62	(a)
3	(b)	23	(b)	43	(a)	63	(b)
4	(b)	24	(a)	44	(b)	64	(a)
5	(c)	25	(b)	45	(a)	65	(a)
6	(d)	26	(a)	46	(d)	66	(c)
7	(d)	27	(a)	47	(c)	67	(d)
8	(a)	28	(c)	48	(d)	68	(a)
9	(d)	29	(a)	49	(a)	69	(a)
10	(d)	30	(a)	50	(b)	70	(a)
11	(c)	31	(a)	51	(b)	71	(a)
12	(a)	32	(a)	52	(c)	72	(a)
13	(a)	33	(d)	53	(b)	73	(a)
14	(b)	34	(c)	54	(c)	74	(d)
15	(a)	35	(d)	55	(c)	75	(a)
16	(b)	36	(a)	56	(a)	76	(a)
17	(c)	37	(a)	57	(a)	77	(a)
18	(a)	38	(c)	58	(a)	78	(a)
19	(c)	39	(c)	59	(d)	79	(d)
20	(b)	40	(a)	60	(c)	80	(c)

81	(b)	108	(b)	135	(d)	162	(d)
82	(d)	109	(a)	136	(b)	163	(c)
83	(c)	110	(a)	137	(b)	164	(b)
84	(b)	111	(a)	138	(c)	165	(a)
85	(c)	112	(b)	139	(b)	166	(b)
86	(d)	113	(b)	140	(c)	167	(b)
87	(b)	114	(a)	141	(b)	168	(b)
88	(d)	115	(a)	142	(d)	169	(d)
89	(b)	116	(b)	143	(d)	170	(d)
90	(b)	117	(d)	144	(b)	171	(b)
91	(c)	118	(d)	145	(d)	172	(b)
92	(d)	119	(c)	146	(b)	173	(a)
93	(a)	120	(a)	147	(b)	174	(d)
94	(b)	121	(d)	148	(a)	175	(b)
95	(a)	122	(d)	149	(d)	176	(a)
96	(d)	123	(d)	150	(d)	177	(d)
97	(b)	124	(a)	151	(a)	178	(b)
98	(a)	125	(d)	152	(d)	179	(d)
99	(c)	126	(b)	153	(b)	180	(b)
100	(d)	127	(d)	154	(d)	181	(a)
101	(c)	128	(d)	155	(c)	182	(b)
102	(c)	129	(c)	156	(d)	183	(c)
103	(c)	130	(a)	157	(a)	184	(c)
104	(c)	131	(d)	158	(b)	185	(b)
105	(d)	132	(d)	159	(a)	186	(d)
106	(b)	133	(b)	160	(d)	187	(c)
107	(a)	134	(d)	161	(c)	188	(c)

Fill in the Blanks

1. Braconidae
2. Bethylidae
3. 50,000 - 1,00,000
4. Wider
5. Potato Tuber Moth

Match the following

1. a) - ii, b) - i, c) - iv, d) - iii
2. a) - iv, b) - iii, c) - ii, d) - i
3. a) - iii, b) - ii, c) - iv, d) - i
4. a) - iv, b) - ii, c) - iii, d) - i
5. a) - i, b) - iii, c) - ii, d) - iv
6. a) - iv, b) - ii, c) - i, d) - iii
7. a) - ii, b) - i, c) - iv, d) - iii

Chapter 6

Chemical Control and Toxicology

Multiple Choice Questions

1. **Institute of pesticide formulation technology is located at**
   (*a*) Gurgaon (*b*) New Delhi
   (*c*) Faridabad (*d*) Chandigarh

2. **Which class of chemical insecticides is characterized by a relatively high degree of environmental persistence?**
   (*a*) Carbamates (*b*) Organophosphates
   (*c*) Synthetic pyrethroids (*d*) Chlorinated hydrocarbons

3. **Organophosphate and Carbamate insecticides work by disrupting an insects**
   (*a*) Digestive system (*b*) Nervous system
   (*c*) Respiratory system (*d*) Endocrine system

4. **The Central Insecticides Laboratory (CIL) is located at**
   (*a*) Gurgaon (*b*) New Delhi
   (*c*) Faridabad (*d*) Chandigarh

5. **Mode of action of spinosad**
   (*a*) Blocks nicotinic acetylcholine receptor (*b*) Affects GABA receptor
   (*c*) Both (*a*) & (*b*) (*d*) None of these

6. **An insecticide having low LC_{50} value indicates**
   (*a*) More toxic (*b*) Less toxic
   (*c*) Non toxic (*d*) Moderately toxic

7. IARI and NII develops a male contraceptive (spermaticidal) from neem which is known as

(*a*) Neembidin (*b*) Neembicidin

(*c*) Praneem or NJM - 76 (*d*) Nemalok

8. Carbamates discovered in the year

(*a*) 1932 (*b*) 1947

(*c*) 1952 (*d*) 1941

9. Among two regional pesticide testing laboratories (RPTLS) one situated at Kanpur and another at

(*a*) Gurgaon (*b*) New Delhi

(*c*) Faridabad (*d*) Chandigarh

10. Which colour on the level of the container indicates the highly toxic pesticide

(*a*) Bright Red (*b*) Bright yellow

(*c*) Bright Blue (*d*) Bright Green

11. Endrin is the__________isomer of dieldrin

(*a*) Stereo (*b*) Endo-endo

(*c*) Endo-exo (*d*) None of these

12. The insecticide act, 1968 were implemented in-

(*a*) 1968 (*b*) 1971

(*c*) 1986 (*d*) 1917

13. Neem (*Azadirachta indica*) content insecticidal property chemical belongs to the group

(*a*) Tri-Cyclic Acid (*b*) Cholesterol

(*c*) Tetrano-tri-terpinoide (*d*) Monosaccharide

14. Expand MRL

(*a*) Maximum Reproduction Level (*b*) Minimum Residue Limit

(*c*) Maximum Residue Limit (*d*) Maximum Residue Level

15. Endosulfan is a

(*a*) Chlorinated hydrocarbon (*b*) Cyclodiene compound

(*c*) Nitroparaffin analogues (*d*) Sulphur compound

16. The oldest insecticide used by humans were of

(*a*) Animal origin (*b*) Plant origin

(*c*) Chemical origin (*d*) Microbial origin

17. The first organic synthetic compound used as insecticide

(*a*) DDT (*b*) Alkayl thiocyanates

(*c*) BHC (*d*) Arsenic

18. Expand DDT

(*a*) Dichloro Diphenyl Trichloroethane

(*b*) Difloro Diphenyl Trichloroethane

(*c*) Dibromo Diphenyl Trichloroethane

(*d*) Trichloro Diphenyl Trichloroethane

19. Which one of the following is a trade name of Endosulfan?

(*a*) Dipel (*b*) Doom

(*c*) Thiodan (*d*) Ekatox

20. Paul Muller was awarded Nobel prize for discovering the insecticidal properties of DDT in the year

(*a*) 1948 (*b*) 1946

(*c*) 1944 (*d*) 1940

21. The period since 1940s has rightly been called the age of

(*a*) Pesticide (*b*) Chemical

(*c*) Botanical (*d*) Technology

22. *Bacillus thuringiensis* is a

(*a*) Contact poison (*b*) Stomach poison

(*c*) Fumigant (*d*) None of these

23. Toxaphene is a

(*a*) Chlorinated terpens (*b*) Cyclodiene compound

(*c*) Nitroparaffin analogues (*d*) DDT analogues

24. The age of pesticide from 1962 to 1976 is known as

(*a*) The era of optimism (*b*) The era of doubt

(*c*) The era of IPM (*d*) None of these

25. "Silent Spring" was wrote by Rachel Carson in the year

(*a*) 1962 (*b*) 1926

(*c*) 1963 (*d*) 1964

26. Aluminium Phosphide is an example of

(*a*) Contact poison (*b*) Stomach poison

(*c*) Fumigant (*d*) None of these

27. Aldrin, Dieldrin are derived by__________reaction.

(a) Tri-Urate (b) Bi-Urate

(c) Diels-Alder (d) Glycosis

28. DDT was synthesized in the year 1874 by

(a) Mular (b) Zeidler

(c) Kogan (d) Needham

29. The developed ability of a strain within an insect species to withstand an insecticide to which it was formerly susceptible is known as

(a) Immunity (b) Resistance

(c) Biotype (d) Resurgence

30. Which of the following is a synthetic derivatives of nereis toxin?

(a) Cartap (b) Bensultap

(c) Thiocyclam (d) All of these

31. DDT is toxic to

(a) Cucurbits (b) Wheat

(c) Rice (d) Both (a) & (b)

32. The recovery of pest populations (sometimes to levels higher than before) following application of an insecticides is known as

(a) Immunity (b) Resistance

(c) Biotype (d) Resurgence

33. Thiocyanates are contact insecticides acting on the insect's__________system.

(a) Digestive (b) Excretory

(c) Circulatory (d) Nervous

34. Which of these insecticides is derived from actinomycetes?

(a) Nereistoxin (b) Dimilin

(c) Plumbago (d) Spinosad

35. Which one of the following is a trade name of Methoxychlor?

(a) Manlate (b) Malachlor

(c) Malaoxone (d) Methochlor

36. Cyclodiene compounds are cyclic, highly chlorinated molecules having an intramolecular endomethylene bridge and prepared by__________reaction.

(a) Tri-Urate (b) Bi-Urate

(c) Diels-Alder (d) Glycosis

37. The respiratory poison effects which stage of respiration

(*a*) Glycolysis (*b*) Kreb's Cycle

(*c*) Electron Transport Chain (*d*) All of these

38. The insecticidal activity of various isomers of DDT falls in the order

(*a*) Para - para > Meta - para > Meta - ortho > Ortho - ortho

(*b*) Meta - para > Para - para > Meta - ortho > Ortho - ortho

(*c*) Meta - ortho > Para - para > Meta - para > Ortho - ortho

(*d*) None of these

39. Kelthane, Dimite, Akar are

(*a*) Insecticide (*b*) Acaricide

(*c*) Fungicide (*d*) Mollacicide

40. Which group of insecticide is the largest in worlds registered insecticide?

(*a*) Organophosphate (*b*) Carbamate

(*c*) Pyradine (*d*) Organo chloride

41. Carbofuran is effectively used against which type of insect

(*a*) Sucking (*b*) Stem borer

(*c*) Bark eating beetle (*d*) Leaf cutter

42. Which of the following is an acaricide (Miticide)?

(*a*) Dimethote (*b*) Monocrotophos

(*c*) Dicofol (Kelthane) (*d*) BHC

43. BHC contain 99 per cent γ isomer termed as

(*a*) HCH (*b*) Marlate

(*c*) Lindane (*d*) Crude BHC

44. Carbofuran is used as toxicant in the form of

(*a*) Solid (*b*) Liquid

(*c*) Fumigant (*d*) Spray

45. The value of lethal dose killed 50 per cent population is called

(*a*) LC_{50} (*b*) LT_{50}

(*c*) LD_{50} (*d*) TL_{50}

46. Strychnine is a poison against

(*a*) Cockroaches (*b*) Crickets

(*c*) Rats (*d*) Birds

47. BHC is a mixture of__________isomers.

(a) 7 (b) 6

(c) 13 (d) 10

48. Recently banned insecticide in India is

(a) Endosulfan (b) Malathion

(c) Carbaryl (d) Diazinon

49. Phorate belongs to

(a) Chlorinated hydrocarbon (b) Organophosphate

(c) Carbamate (d) Dinitrophenol

50. Chemical used against snails is

(a) Formaldehyde (b) Metaldehyde

(c) Benzaldehyde (d) Acetaldehyde

51. DDT was first used in Switzerland in the year

(a) 1941 against Colorado Potato Beetle

(b) 1939 against Diamond Back Moth

(c) 1940 against *Pediculus humanus*

(d) 1941 against Gypsy moth

52. Which of the following insecticide belongs to synthetic pyrethroid group?

(a) Cypermethrin (b) Carbaryl

(c) Monocrotophos (d) Malathion

53. Chemical used to kill snail and slugs are

(a) Acaricides (b) Nematicides

(c) Mollascicides (d) Fungicides

54. The isomer of BHC which is more toxic

(a) Alpha (b) Beta

(c) Gamma (d) Delta

55. The toxicity of DDT is in form of

(a) P-P′-DDT - 65 - 80 per cent and O-P′-DDT - 20 - 25 per cent

(b) P-P′-DDT - 70 - 90 per cent and O-P′-DDT - 15 - 20 per cent

(c) P-P′-DDT - 65 - 80 per cent and O-P′-DDT - 15 - 21 per cent

(d) None of these

56. Which of the following insecticide is systemic in action?

(a) Endosulfan (b) Dichlorovos

(c) Ethion (d) Dimethoate

57. Insecticide is generally formulated as granule

(*a*) BHC (*b*) Phosphamidon

(*c*) Ethion (*d*) Furadon

58. Which of the following is a cyclodiene compound?

(*a*) Endosulfan (*b*) Endrin

(*c*) Heptachlor (*d*) All of these

59. All organochlorines are more or less

(*a*) Soluble in water and insoluble in organic

(*b*) Soluble in organic solvent and insoluble in water

(*c*) Soluble both in organic solvent and water

(*d*) insoluble both in organic solvent and water

60. One of the ingredients used in universal antidote is

(*a*) Acetic acid (*b*) Phenol

(*c*) Charcoal (*d*) Glycine

61. The maximum percentage of insecticide is used in which crop

(*a*) Rice (*b*) Cotton

(*c*) Gram (*d*) Brinjal

62. Following insecticide belongs to orgnophosphate group

(*a*) Monocrotophos (*b*) Carbaryl

(*c*) Endosulfan (*d*) Kelthane

63. DDT is

(*a*) $C_{14}H_{10}Cl_5$ (*b*) $C_{14}H_8Cl_5$

(*c*) $C_{14}H_9Cl_5$ (*d*) $C_2H_2Cl_2$

64. Which of the following is an example of chlorinated hydrocarbon?

(*a*) Aldicarb (*b*) Quinalphos

(*c*) Cypermethrin (*d*) DDT and BHC

65. The correct name of BHC is

(*a*) HCH (*b*) Aldrin

(*c*) LDJ (*d*) Aldicarb

66. Which of the following insecticide has systemic action?

(*a*) Dichlorovos (*b*) Monocrotophos

(*c*) Dimethoate (*d*) All of these

67. Cartap is derived from

(a) *Sacaropolyspora spinosa* (b) *Bacillus thuringiensis*

(c) *Lumbrineris heteropoda* (d) None of these

68. Most important enzyme of MFO system is

(a) Phosphate (b) Pechlorinase

(c) Carboxyesterase (d) Cytochrome P - 450

69. Which type of toxic creates widespread ecological problem?

(a) Acute toxicity (b) Dermal toxicity

(c) Mammalian toxicity (d) Chronic toxicity

70. An insecticide which is used as seed treatment for management of sucking insect pests

(a) Imidachloprid (b) Endosulfan

(c) Decamethrin (d) Fenvalerate

71. Neem referred as

(a) IGR (b) AchE inhibitor

(c) GABA activator (d) None of these

72. Which of the following is an Acaricide?

(a) Pirmicarb (b) Dicofol

(c) Profenphos (d) Sevidol

73. Which of the following is a trade name of aldicarb?

(a) Rogor (b) Sevin

(c) Temic (d) Carbaryl

74. An insecticide which is having both fumigant and penetrate action

(a) Monocrotophos (b) DDVP

(c) Chlorpyriphos (d) Phosphamidon

75. Azadiractin have__________stereo isomers.

(a) 7 (b) 8

(c) 9 (d) 10

76. Which of the following insecticide is effective against termite?

(a) Nimbecidine (b) Endosulfan

(c) Dimethoate (d) Chlorpyriphos

77. Neem based insecticidal product is

(a) Nimbecidine (b) Endosulfan

(c) Thiodan (d) Chlorpyriphos

78. Which of the following is not a carbamate insecticide?

(*a*) Aldicarb (*b*) Carbofuran

(*c*) Dimethoate (*d*) Methomyl

79. Salannin is derived from

(*a*) Tobacco (*b*) Neem

(*c*) Derris (*d*) Quassia

80. Which of the following is an example of carbamate insecticide?

(*a*) Cypermethrin (*b*) Bioneem

(*c*) Aldicarb (*d*) Dipel

81. Insecticide used for seed treatment against sucking pests

(*a*) Endrin (*b*) Diazinon

(*c*) Imidachloprid (*d*) Malathion

82. The insecticide which is used as soil application

(*a*) Phorate (*b*) Carbofuran

(*c*) Chlorpyriphos (*d*) All of these

83. The family of *Sabadilla officinale* is

(*a*) Compositae (*b*) Liliaceae

(*c*) Solanaceae (*d*) Anacardiace

84. Dimethoate belongs to

(*a*) Fumigant (*b*) Stomach poison

(*c*) Systemic insecticide (*d*) Contact insecticide

85. Maximum use of pesticide is in which vegetable

(*a*) Brinjal (*b*) Okra

(*c*) Tomato (*d*) Potato

86. The insecticide which is being opposed by environmentalist is

(*a*) Sulphur (*b*) DDVP

(*c*) Chlorpyriphos (*d*) Imidachloprid

87. Rotenone is derived from

(*a*) *Derris* spp. (*b*) *Sabadilla* spp.

(*c*) *Tagetas* spp. (*d*) *Sorea* spp.

88. Environmentally opposed insecticide is

(*a*) Chlorpyriphos (*b*) DDVP

(*c*) Sulphur (*d*) Endosulfan

89. Insecticide which is recommended for the use in vegetable

(*a*) Malathion (*b*) Aldicarb
(*c*) Cypermethrin (*d*) Phosphamidan

90. The phenomenon of insect resistance to one type of insecticide, providing resistance to other insecticides with similar modes of action

(*a*) Multiple resistance (*b*) Cross resistance
(*c*) Oligo resistance (*d*) None of these

91. The household brand, mixture of DDT, Pyrethrum in solvent oil is known as

(*a*) Sol (*b*) Flit
(*c*) Gel (*d*) Emerson

92. Widely used acaricide against mite is

(*a*) DDVP (*b*) Sulphur
(*c*) Kelthane (*d*) Monocrotophos

93. Plant origin pesticide is

(*a*) Sulphur (*b*) Carbaryl
(*c*) Aldicarb (*d*) NSKE

94. The irreversible inversion of acetyl cholinesterase is caused by

(*a*) Carbofuran (*b*) BHC
(*c*) Carbaryl (*d*) Malathion

95. Pyrethrins are derived from which part of *Crysanthemum einerariaefolium*

(*a*) Stem (*b*) Dried flower
(*c*) Leaf (*d*) Root

96. Example of insecticide having long residual effect

(*a*) Aldicarb (*b*) Phorate
(*c*) DDT (*d*) Cypermethrin

97. Example of insecticide which belongs to organophosphate group

(*a*) Sulphur (*b*) Quinalphos
(*c*) Endosulfan (*d*) Cypermethrin

98. The pyrethroid which is used in mosquito coils

(*a*) Allethrin (*b*) Permethrin
(*c*) Cypermethrin (*d*) None of these

99. Which of the following is an insecticide of animal origin?

(*a*) Strychnine (*b*) Padan
(*c*) Ro-Dex (*d*) Certox

100. Insecticide used for mosquito killing

(*a*) Endosulfan (*b*) Malathion
(*c*) Allethrin (*d*) Diazinone

101. Name of effective anticoagulant used for rat control

(*a*) Al_2PO_4 (*b*) Warfarin
(*c*) Hydrogen cyanide (*d*) Zinc phosphide

102. First systemic inorganic insecticide

(*a*) Sodium carbonate (*b*) Sodium oxalate
(*c*) Sodium selenate (*d*) Sodium bicarbonate

103. Nicotine is

(*a*) $C_{10}H_{14}N_2$ (*b*) $C_{10}H_{12}N_4$
(*c*) $C_{12}H_{10}N_2$ (*d*) $C_8H_{12}N_{10}$

104. For effective of aphid and white flies used

(*a*) Contact insecticide (*b*) Fumigants
(*c*) Systemic insecticide (*d*) Nerve poison

105. Insecticide having high LD_{50} value indicates

(*a*) Highly toxic (*b*) Neutral
(*c*) Less toxic (*d*) None of these

106. Which of these insecticides belongs to the neonicotinoid group?

(*a*) Thiamethoxam (*b*) Imidachloprid
(*c*) Acetamiprid (*d*) All of these

107. Thallium sulphate is used for

(*a*) Sweet loving ant (*b*) Protein loving ant
(*c*) Cockroach (*d*) Rats and mice

108. Which is used as effective sticker?

(*a*) Coaltar (*b*) Teepol
(*c*) Phenol (*d*) Neem oil

109. Insecticide having anticholinesterase in action

(*a*) Lindane (*b*) Aldrin
(*c*) Phorate (*d*) Malathion

110. Which of these is anti-coagulant for rats?

(*a*) Zinc phosphide (*b*) Alluminium phosphide
(*c*) Strychnine hydrochloride (*d*) Warfarin

111. Borax is

(a) Sodium tetra borate (b) Sodium tri borate
(c) Sodium borate (d) Sodium hexa borate

112. Extremely toxic poison indicate through

(a) Bright yellow (b) Bright red
(c) Bright green (d) Bright blue

113. Atropine is used as

(a) Antidote (b) Attractant
(c) Antifeedant (d) Allomone

114. Celphos tablets are used for the management of

(a) Stored product pests (b) Field pests
(c) House hold pests (d) None of these

115. Which can be used as insecticide, fungicide as well as Acaricide?

(a) Fluorides (b) Arsenites
(c) Sulphur (d) Phosphorus

116. Which of the following is not an example of biopesticide?

(a) Carbaryl (b) Bioneem
(c) Biolep (d) Dipel

117. Which of the following insecticide having fumigant action?

(a) Malathion (b) DDVP
(c) Cypermethrin (d) Lindane

118. Which of the following is a fumigant?

(a) Methyl bromide (b) Ethylene dichloride
(c) Chloropicrin (d) All of these

119. Cryolite is

(a) NaF (b) Na_2SiF_6
(c) Na_2AlF_6 (d) $NaAlF_6$

120. Which of the following improve the emulsifying, wetting and spreading properties of the mixture?

(a) Diluents (b) Stickers
(c) Surfactants (d) Deodorants

121. White grub effectively controlled by which of the following chemical

(a) Endosulfan (b) Phorate
(c) Aldicarb (d) Chlorpyriphos

122. Ethylene dichloride is generally mixed with carbon tetrachloride in the ratio of

(*a*) 1 : 3 (*b*) 3 : 1

(*c*) 1 : 2 (*d*) 2 : 1

123. Actually paris green is

(*a*) Copper oxychloride (*b*) Copper acetoarsenate

(*c*) Copper sulphate (*d*) Cupric oxide

124. Celphos release a poisonous gas namely

(*a*) Hydrogen gas (*b*) Phosphine gas

(*c*) Sulphur dioxide (*d*) None of these

125. Malathion belongs to the group

(*a*) Carbamate (*b*) Organophosphate

(*c*) Organochlorine (*d*) Synthetic pyrethroids

126. The fumigant which is widely used in quarantine stations at entry points is

(*a*) Methyl bromide (*b*) Ethylene dichloride

(*c*) Chloropicrin (*d*) None of these

127. Paris green is a

(*a*) Contact poison (*b*) Stomach poison

(*c*) Protoplasmic poison (*d*) Respiratory poison

128. Nephthaline is used as

(*a*) Fumigant (*b*) Soil insecticide

(*c*) Fungicide (*d*) Nerve poison

129. Which is not a botanical insecticide?

(*a*) Malathion (*b*) Pyrethrum

(*c*) NSKE (*d*) Nicotene

130. Which of the following fumigant will be completely phased out in developing countries by 2015 by the 1992 Montreal Convention?

(*a*) Methyl bromide (*b*) Ethylene dichloride

(*c*) Chloropicrin (*d*) None of these

131. Lead arsenate was first used in the year

(*a*) 1882 (*b*) 1892

(*c*) 1890 (*d*) 1872

132. Insecticide of plant origin are

(a) Nicotine
(b) Achook
(c) Pyrethrum
(d) All of these

133. Full form of BHC is

(a) Benzene Hexa Chloride
(b) Benzo Hexane Chloride
(c) Bromo Hydro Chloride
(d) All of these

134. Which of the following fumigant depletes the ozone layer?

(a) Methyl bromide
(b) Ethylene dichloride
(c) Chloropicrin
(d) Phosphine

135. On the container of insecticide bright blue colour denotes

(a) Extremely toxic
(b) Moderately toxic
(c) Highly toxic
(d) Slightly toxic

136. Fumigants are available as

(a) Only liquid
(b) Only solid
(c) Both liquid and solid
(d) Only gaseous state

137. Which of the following insecticide is banned for use in food grain?

(a) Celphos
(b) EDB
(c) Malathion
(d) None of these

138. Sucking insect pests are effectively controlled by

(a) Systemic insecticide
(b) Stomach poisons
(c) Fumigants
(d) Contact insecticides

139. The antidote of Nicotine is

(a) 2 PAM
(b) Atropine sulphate
(c) Phenobarbital
(d) Potassium Permanganate solution

140. Scale insect of citrus is controlled by

(a) BHC
(b) DDT
(c) Endosulfan
(d) Hydrocarbon oil

141. Inorganic salt is toxic to insect by

(a) Ingestion
(b) Digestion
(c) Contact
(d) None of these

142. The crop which consumes highest pesticides in India is

(*a*) Rice (*b*) Groundnut

(*c*) Cotton (*d*) Sugarcane

143. In universal antidote the ratio of activated charcoal, MgO and tannic acid is

(*a*) 2 : 0.5 : 1 (*b*) 2 : 1 : 1

(*c*) 1 : 1 : 1 (*d*) 3 : 1 : 1

144. The ratio of zinc phosphate in poison bait for rat is

(*a*) 1 per cent (*b*) 2 per cent

(*c*) 3 per cent (*d*) 4 per cent

145. The fumigation of rat burrows is done with

(*a*) EDB (*b*) Methyl bromide

(*c*) Atropine (*d*) Aluminum phosphide

146. The highest consumer of pesticides per unit area is

(*a*) India (*b*) Taiwan

(*c*) Japan (*d*) USA

147. PCP is widely used insecticide for termite control

(*a*) Para chloro phenol (*b*) Penta chlorophloro phenol

(*c*) Pest calling path (*d*) None of these

148. Carbon disulphide is used to control of

(*a*) Insects (*b*) Fungus

(*c*) Bacteria (*d*) Nematodes

149. Most effective chemical against cutworm is

(*a*) Methyl Parathion (*b*) Endosulfan

(*c*) Chlorpyriphos (*d*) DDVP

150. Following insecticide is safely recommended on vegetables

(*a*) Carbaryl (*b*) Aldicarb

(*c*) Endrin (*d*) Malathion

151. Muscarinic symptoms produced by accumulation of Ach at

(*a*) Neuromascular junction

(*b*) Neuroeffector junction

(*c*) Neuro hemal organs

(*d*) Between motor nerves and nerves and muscles

152. Rodenticides are used to kill

(*a*) Mites (*b*) Insects

(*c*) Rat and mice (*d*) Fungus

153. Most effective control of cotton jassid is

(*a*) BHC (*b*) Copper sulphate

(*c*) Methyl Parathion (*d*) All of these

154. Number of pesticides including Bio-pesticides formulations registered in India

(*a*) 170 (*b*) 200

(*c*) 221 (*d*) 300

155. Organo phosphate and carbamates inhibit the action of

(*a*) Ach (*b*) AchE

(*c*) Adrinalin (*d*) None of these

156. Celphos used for the control of

(*a*) Storage pests (*b*) Coconut mite

(*c*) Sugarcane mite (*d*) Scale insect

157. Which insecticide has contact and fumigant action

(*a*) Endosulfan (*b*) Diazinon

(*c*) DDVP (*d*) Phosphamidon

158. The insecticide used as stem injection or root feeding recommended for management of coconut black headed caterpillar is

(*a*) Chlorpyriphos (*b*) Monocrotophos

(*c*) Carbofuran (*d*) Malathion

159. TLm is termed as

(*a*) Tolerance limit mean (*b*) Median tolerance limit

(*c*) Total limit mean (*d*) Median total limit

160. Which of the following insecticide is stored in the human fat

(*a*) BHC (*b*) DDT

(*c*) EDB (*d*) DDVP

161. Example of carcinogenic group of insecticide are

(*a*) Organochlorine (*b*) Organophosphate

(*c*) Carbamate (*d*) Synthetic pyrethroids

162. Which of the following is used as a sticker along with insecticide for better spread?

(*a*) Teepol (*b*) Alcohol

(*c*) Phenol (*d*) Charcoal

163. LD_{50} value is measured in

(*a*) mg/kg (*b*) ppm

(*c*) mg/gm (*d*) gm/kg

164. Which insecticide has maximum residual effect?

(*a*) DDT (*b*) Phorate

(*c*) Endosulfan (*d*) Malathion

165. Name of anticoagulant which is used as rodenticide

(*a*) Zinc phosphide (*b*) Warfarin

(*c*) Parathion (*d*) EDB

166. Insecticide which is effective against termites

(*a*) Endosulfan (*b*) Monocrotophos

(*c*) Metasystox (*d*) Chlorpyriphos

167. Time required for half of its residues to react and kill 50 per cent of the insect population of an insecticide is known as

(*a*) LD_{50} (*b*) RL_{50}

(*c*) LC_{50} (*d*) MD_{50}

168. Indiscriminate use of pesticide and cultivation of high yielding variety often leads to

(*a*) Destruction of natural enemies

(*b*) Insect races with pesticide resistance

(*c*) Secondary pest out breaks

(*d*) All of these

169. Insecticide approved for aerial spray

(*a*) Lindane 20 EC (*b*) Malathion 50 EC

(*c*) Phosphamidon 85 WSC (*d*) All of these

170. Which of the following is not a Bio-pesticide?

(*a*) Biobit (*b*) Dipel

(*c*) Phosalone (*d*) Biolep

171. Insecticide residue is calculated in

(*a*) ppm (*b*) gm/kg

(*c*) µg/gm (*d*) All of these

172. Resistance of insecticide is developed due to

(*a*) Storage of toxicant within body

(*b*) Lower concentration of pesticide use

(*c*) Conversion of toxicant to non toxic metabolites

(*d*) All of these

173. Diflubenzuron is a derivative of

(*a*) Difloro compounds (*b*) Urea

(*c*) Phosphoric acid (*d*) None of these

174. Atropine is a/an

(*a*) Antidote (*b*) Antifeedant

(*c*) Repellent (*d*) None of these

175. Which of the following is a non inflammable gas?

(*a*) CCl_4 (*b*) HCN

(*c*) Carbon tetrachloride (*d*) Both (*a*) & (*b*)

176. If one gram of pesticide formulation containing 50 per cent active ingredient is mixed with 01 liter of water what will be concentration of the active ingredient in the spray fluid

(*a*) 5.00 ppm (*b*) 50 ppm

(*c*) 500 ppm (*d*) 5000 ppm

177. Pesticide is suitable for controlling the white grubs

(*a*) Quinalphos (*b*) Chlorpyriphos

(*c*) Phorate (*d*) All of these

178. Universal antidote contains

(*a*) 7 parts of activated charcoal + 3.5 parts MgO + 3.5 parts Tannic acid

(*b*) 6 parts of activated charcoal + 3 parts MgO + 7 parts Tannic acid

(*c*) 8 parts of activated charcoal + 4.5 parts MgO + 4 parts Tannic acid

(*d*) 6 parts of activated charcoal + 3.5 parts MgO + 4.5 parts Tannic acid

179. Generally ULV use

(*a*) 5 - 400 lit/ha (*b*) More than 400 lit/ha

(*c*) Less than 5 lit/ha (*d*) None of these

180. The minimum amount of toxicant required to kill 50 percent of the population of test animal is called

(*a*) LD_{50} (*b*) RL_{50}

(*c*) LC_{50} (*d*) MD_{50}

181. The level of a chemical residue to which daily exposure over a life time is not thought to cause appreciable risk is called

(*a*) Accordable Daily Intake
(*b*) Acceptable Daily Intake
(*c*) Appreciable Daily Intake
(*d*) None of these

182. Which of the following is a synergists?

(*a*) Piperonyl butoxide
(*b*) Sesamin
(*c*) Sesamolin
(*d*) All of these

183. The common range of toxicant in WDP is

(*a*) 25 - 50 per cent
(*b*) 20 - 70 per cent
(*c*) 30 - 32 per cent
(*d*) None of these

184. Insecticidal properties of DDT were first discovered by__________in 1939.

(*a*) Paul Mullar
(*b*) Faradey
(*c*) Zaidler
(*d*) Von

185. The measurement of the potency of a stimulus by means of reactions it produces in the living matter is called

(*a*) Residue
(*b*) Bioassay
(*c*) Deposition
(*d*) Magnification

186. The insecticide that is safe to bees is

(*a*) Phorate
(*b*) Endosulfan
(*c*) Monocrotophos
(*d*) Acephate

187. A pesticide used in ULV application

(*a*) Malathion
(*b*) DDT
(*c*) BHC
(*d*) All of these

188. The optimum safe concentration of the left over pesticide or its degradation products according to good agricultural practices is known as

(*a*) Maximum Reproduction Level
(*b*) Minimum Residue Limit
(*c*) Maximum Residue Limit
(*d*) Maximum Residue Level

189. Which of these insecticides is originated from a marine annelid, *Lumbrineris heteropoda*?

(*a*) Nereistoxin
(*b*) Dimilin
(*c*) Plumbago
(*d*) Spinosad

190. The insecticide which is formulated as 100 per cent EC

(*a*) DDVP (Dichlorovos)
(*b*) Monocrotophos
(*c*) Dipel
(*d*) Acephate

191. Toxicant percentage of granule is

(a) 2 - 10 per cent
(b) 1 - 20 per cent
(c) 0.25 - 2.4 per cent
(d) 0.1 - 25 per cent

192. BHC was first synthesized by__________in 1825.

(a) Paul Mullar
(b) Faradey
(c) Zaidler
(d) Von

193. Insecticides which are absorbed by the plant and translocated in the system are called

(a) Stomach insecticides
(b) Contact insecticides
(c) Systemic insecticides
(d) None of these

194. The total number of pesticides restricted for use in India is

(a) 13
(b) 5
(c) 21
(d) 25

195. Dust particle ranges from

(a) 1 - 40 μ
(b) 2 - 20 μ
(c) 0.25 - 2.4 mm
(d) 0.1 - 25 mm

196. Which of the following first synthetic pyrethrum analogue was commercially introduced in 1949?

(a) Cypermethrin
(b) Permethrin
(c) Allethrin
(d) None of these

197. The toxic gas that is released when Aluminium phosphide react with moisture is

(a) Ammonia
(b) Aluminum phosphate
(c) Phosphine
(d) None of these

198. Rotenone is obtained from which part of *Derris elliptica*

(a) Roots
(b) Stem
(c) Leaf
(d) Flowers

199. Which of the following is pseudosynergist?

(a) Oleic acid
(b) Lubricating oil
(c) Oil part of pine/sesame
(d) All of these

200. A situation where a population after having been suppressed, rebounds to numbers greater than before suppression occurred is known as

(a) Biotype
(b) Resistance
(c) Resurgence
(d) Ecotype

201. WDP are more__________than dust.

(*a*) Concentrated (*b*) Diluted

(*c*) Thinned (*d*) None of these

202. Thallium sulphate (Ti_2SO_4) is mixed with a sweet substance and used as a__________for ants.

(*a*) Pheromone trap (*b*) Poison bait

(*c*) Food (*d*) None of these

203. The adverse effect of insecticide on plants due to the pesticide treatment is known as

(*a*) Residue (*b*) Phytotoxic

(*c*) Toxicity (*d*) None of these

204. The first report of insecticide resistance in India was reported in the insect

(*a*) *Plutella xylostella* (*b*) Singhara beetle

(*c*) Gypsy moth (*d*) Gram pod borer

205. The insect which has developed resistance to maximum number of insecticides under field conditions is

(*a*) *Plutella xylostella* (*b*) Singhara beetle

(*c*) Gypsy moth (*d*) Gram pod borer

206. For an insecticide to enter through spiracles it should have

(*a*) High vapour pressure (*b*) Low vapour pressure

(*c*) No vapour pressure (*d*) None of these

207. IN 930 and Piperonyl butoxide are example of

(*a*) Synergist (*b*) Antagonist

(*c*) Activator (*d*) Both (*a*) & (*c*)

208. Malathion is converted into Malaoxon, which is an efficient inhibitor of

(*a*) Cholinesterase (*b*) Acetyl choline

(*c*) Choline acetylase (*d*) None of these

209. A phenomenon where there is increase in the activity of a compound by addition of non toxic compound which produces effect that is greater than the sum of their individual is called

(*a*) Synergism (*b*) Physiological selectivity

(*c*) Insecticide Resistance (*d*) Resurrection

210. Which of the following is characteristic property of a contact insecticide?

(*a*) A polar (*b*) Lipophilic

(*c*) Low partition coefficient (*d*) All of these

211. Adjuvants are discovered in the year

(*a*) 1939 (*b*) 1938

(*c*) 1947 (*d*) 1940

212. When two organisms are exposed to a toxicant but one is killed and the other is survived by some physiological or biochemical mechanism, tolerating the exposure is called

(*a*) Synergism (*b*) Physiological selectivity

(*c*) Insecticide Resistance (*d*) Resurrection

213. The ability of insects to tolerate doses of toxicant which would prove lethal to the majority of individuals in a normal population of the same species is called

(*a*) Synergism (*b*) Physiological selectivity

(*c*) Resistance (*d*) Resurrection

214. The resistance in insects is due to the following physiological consideration

(*a*) Conversion of toxicant to non toxicant metabolites

(*b*) Excretion of toxicant

(*c*) Storage of toxicant

(*d*) All of these

215. Iso propyl cresols is a

(*a*) Stabilizer (*b*) Softener

(*c*) Carrier (*d*) Sticker

216. In housefly the resistance to DDT is due to recessive gene called kdr (knock down resistance) on chromosome

(*a*) 3 (*b*) 2

(*c*) 5 (*d*) 4

217. Nicotine is metabolized to__________in plants, animals and man.

(*a*) Colchicines (*b*) Cotinine

(*c*) Niacin (*d*) Choline

218. In DDT metabolism which one of the following is true

(*a*) DDT is oxidized to Dichlorodiphenyl acetic acid (DDA) in rats

(*b*) DDT is oxidized to Dicofol in *Drosophila*

(*c*) DDT is oxidized to Dichloro benzophenone (DBP) in cockroaches

(*d*) All of these

219. Which of the following is used to reduce the phytotoxicity of insecticides?

(*a*) Stabilizer (*b*) Softener

(*c*) Marker (*d*) Sticker

220. Reproductive stimulation by sublethal doses of pesticides is known as

(*a*) Hormoligosis (*b*) Reflex bleeding

(*c*) Physiological selectivity (*d*) None of these

221. Abamectin is a natural fermentation product obtained from *Streptomyces avermitilis* is a

(*a*) Soil bacterium (*b*) Soil actinomycetes

(*c*) Aquatic fungus (*d*) Marine annelid

222. In resistant insects and house flies DDT is converted into

(*a*) Dichlorodiphenyl acetic acid (DDA)

(*b*) Dichloro benzophenone (DBP)

(*c*) Dichlorodiphenyl dichloroethane (DDD)

(*d*) Dichlorodiphenyl dichloroethylene (DDE)

223. Glue, blood albumin, casein, resin are

(*a*) Sticker (*b*) Emulsifiers

(*c*) Spreaders (*d*) Stabilizer

224. Emamectin benzoate is a novel semi-synthetic derivative of the natural product

(*a*) Abamectin (*b*) Avermectin

(*c*) Ivermectin (*d*) None of these

225. Indoxacarb which is a member of oxadiazine class of insecticides acts by inhibiting__________ions entry into nerve cells.

(*a*) Potassium (*b*) Sodium

(*c*) Calcium (*d*) None of these

226. Spinosad is a natural class of insecticide produced by an actinomycete, under natural fermentation

(*a*) *Sacaropolyspora spinosa* (*b*) *Bacillus thuringiensis*

(*c*) *Lumbrineris heteropoda* (*d*) None of these

227. Compared to first generation pesticides, the newer second and third generation compounds are

(*a*) More selective and less persistent (*b*) More toxic and less selective

(*c*) More persistent and less selective (*d*) None of these

228. What is the generally accepted mode of action for organophosphate insecticides?

(*a*) Chitin inhibitor (*b*) Stomach poison

(*c*) Cholinesterase inhibitor (*d*) Respiratory toxin

229. Spinosad is a

(*a*) Contact poison (*b*) Stomach poison

(*c*) Both (*a*) & (*b*) (*d*) None of these

230. Example of dust additives is

(*a*) Calcium oxide (*b*) Clays

(*c*) Attapulgites (*d*) All of these

Fill in the Blanks

1. Those insecticides which enter the body of the insects by penetrating through cuticle is called as__________.

2. Those insecticides which enter the body of the insects through its food is called as__________.

3. The toxicants which kill the insect by destruction of cellular protoplasm of the mid-gut epithelium are called__________.

4. __________is the daily dose of a chemical which during an entire life-time, appears to be without appreciable risk on the basis of all facts known at that time.

5. Acceptable Daily Intake is expressed in__________.

6. When acetylcholine reacts with cholinesterase an enzyme complex is formed, commonly known as__________.

7. When an insecticide is selective in its action by creating more toxicity to a particular group of organism is termed as__________.

8. A very useful index of selectivity is the VSR, stands for__________.

9. The development of an ability in a strain of insects to tolerate doses of toxicant which would prove lethal to the majority of individuals in a normal population of the same species is known as__________.

10. The amount of initially laid down chemical insecticide on the surface is known as__________.

11. The amount of insecticide left over after a lapse of time is known as__________.

12. The accumulation of insecticides in various biological system is called__________.

13. Strychnine derived from *Nux vomica* are used as__________.

14. __________is the process to determine the amount of unsaturated hydrocarbons.

15. Pearson's square method is adopted for diluting__________.

Match the Following

1. Match the following insecticides with their mode of action

(a) Inert dusts	i) Physical poison
(b) Aluminium phosphide	ii) Stomach poison
(c) Chlorpyriphos	iii) Nerve poison
(d) *Bacillus thuringiensis*	iv) Respiratory poison

2. Match the common name of insecticides with their trade names

(a) Aluminium phosphide	i) Celphos
(b) Endosulfan	ii) Dursban
(c) Chlorpyriphos	iii) Thiodan
(d) Monocrotophos	iv) Nuvacron

3. Match the insecticide groups with their antidotes

(a) Organophosphates	i) Atrpine sulphate
(b) Carbamates	ii) Atropine
(c) Pyrethroids	iii) Potassium permanganate
(d) Nicotine	iv) Diazepam

4. Match the insecticides with their site of action

(a) Pyrethroids	i) Axonic membrane
(b) Nicotine	ii) Pre synaptic membrane
(c) Cyclodienes	iii) Post synaptic membrane
(d) Organophosphates	iv) Synaptic junction

5. Match the common name of insecticides with their trade names

(a) Abamectin	i) Rekord
(b) Acephate	ii) Starthen
(c) Acetamiprid	iii) Vertimec
(d) Aldicarb	iv) Temic

6. **Match the following bio-agent with their commercial products**

(a) *Bacillus thuringiensis*	i) Bio- Power
(b) *Beauveria bassiana*	ii) Biolep
(c) *Metarhizium anisopliae*	iii) Bioshield
(d) *Pseudomonas fluorescens*	iv) Bio Magic

7. **Match the following insecticides with their group**

(a) Propagate	i) Carbamate
(b) Teflubenzuron	ii) Synthetic Pyrethroides
(c) Propoxur	iii) Acaricide
(d) Bifenthrin	iv) Insect Growth Regulator

8. **Match the following pesticides with their generations as has been grouped**

(a) Petroleum products	i) Fourth generation
(b) Dinitro compounds	ii) Third generation
(c) Juvenile hormone	iii) Second generation
(d) Precocene	iv) First generation

9. **Match the following organic insecticides with their chemical composition**

(a) Nicotine	i) $C_{21}H_{30}O_3$
(b) Pyrethroids	ii) $C_{10}H_{14}N_2$
(c) Rotenoids	iii) $C_{23}H_{22}O_6$
(d) Cevadine	iv) $C_{32}H_{49}N$

10. **Match the insecticide with the poisoning symptoms they produced in insects**

(a) DDT	i) Jitters
(b) BHC	ii) Fanning movements
(c) Rotenone	iii) Restlessness
(d) Parathion	iv) Inactiveness

Answers

Multiple Choice Questions

1	(a)	21	(a)	41	(b)	61	(b)
2	(d)	22	(b)	42	(c)	62	(a)
3	(b)	23	(a)	43	(c)	63	(c)
4	(c)	24	(b)	44	(a)	64	(d)
5	(c)	25	(b)	45	(c)	65	(a)
6	(a)	26	(c)	46	(c)	66	(d)
7	(c)	27	(c)	47	(b)	67	(c)
8	(b)	28	(b)	48	(a)	68	(d)
9	(d)	29	(b)	49	(b)	69	(d)
10	(b)	30	(d)	50	(b)	70	(a)
11	(b)	31	(a)	51	(a)	71	(a)
12	(b)	32	(d)	52	(a)	72	(b)
13	(c)	33	(d)	53	(c)	73	(c)
14	(c)	34	(d)	54	(c)	74	(b)
15	(a)	35	(a)	55	(c)	75	(a)
16	(b)	36	(c)	56	(d)	76	(d)
17	(b)	37	(d)	57	(d)	77	(a)
18	(a)	38	(a)	58	(d)	78	(c)
19	(c)	39	(b)	59	(b)	79	(b)
20	(a)	40	(a)	60	(c)	80	(c)

81	(c)	115	(c)	149	(c)	183	(a)
82	(d)	116	(a)	150	(d)	184	(a)
83	(b)	117	(b)	151	(b)	185	(b)
84	(c)	118	(d)	152	(c)	186	(b)
85	(a)	119	(c)	153	(c)	187	(a)
86	(b)	120	(c)	154	(c)	188	(c)
87	(a)	121	(b)	155	(b)	189	(a)
88	(b)	122	(b)	156	(a)	190	(a)
89	(a)	123	(b)	157	(c)	191	(a)
90	(b)	124	(b)	158	(b)	192	(b)
91	(b)	125	(b)	159	(b)	193	(c)
92	(c)	126	(a)	160	(b)	194	(a)
93	(d)	127	(b)	161	(d)	195	(a)
94	(d)	128	(a)	162	(a)	196	(c)
95	(b)	129	(a)	163	(a)	197	(c)
96	(c)	130	(a)	164	(a)	198	(a)
97	(b)	131	(b)	165	(b)	199	(d)
98	(a)	132	(d)	166	(d)	200	(c)
99	(b)	133	(a)	167	(b)	201	(a)
100	(c)	134	(a)	168	(d)	202	(b)
101	(b)	135	(b)	169	(b)	203	(b)
102	(c)	136	(c)	170	(c)	204	(b)
103	(a)	137	(b)	171	(a)	205	(a)
104	(c)	138	(a)	172	(d)	206	(a)
105	(c)	139	(d)	173	(b)	207	(a)
106	(d)	140	(d)	174	(a)	208	(a)
107	(a)	141	(c)	175	(a)	209	(a)
108	(b)	142	(c)	176	(c)	210	(d)
109	(d)	143	(b)	177	(b)	211	(b)
110	(d)	144	(b)	178	(a)	212	(b)
111	(a)	145	(d)	179	(c)	213	(c)
112	(b)	146	(b)	180	(a)	214	(d)
113	(a)	147	(b)	181	(b)	215	(a)
114	(a)	148	(d)	182	(d)	216	(a)

217	(b)	221	(a)	225	(b)	229	(c)
218	(d)	222	(d)	226	(a)	230	(d)
219	(b)	223	(a)	227	(a)		
220	(a)	224	(a)	228	(c)		

Fill in the Blanks

1. Contact poisons
2. Stomach poisons
3. Protoplasmic poisons
4. Acceptable Daily Intake
5. mg/kg body weight/day
6. Michaelis complex
7. Selective toxicity
8. Vertebrate selectivity ratio
9. Resistance
10. Deposit
11. Residue
12. Bioaccumulation
13. Rodenticide
14. Sulphonation
15. Insecticide formulation

Match the Following

1. a) - i, b) - iv, c) - iii, d) - ii
2. a) - i, b) - iii, c) - ii, d) - iv
3. a) - ii, b) - i, c) - iv, d) - iii
4. a) - i, b) - iii, c) - ii, d) - iv
5. a) - iii, b) - ii, c) - i, d) - iv
6. a) - ii, b) - i, c) - iv, d) - iii
7. a) - iii, b) - iv, c) - i, d) - ii

8. **a) - iv, b) - iii, c) - ii, d) - i**

9. **a) - ii, b) - i, c) - iii, d) - iv**

10. **a) - i, b) - ii, c) - iv, d) - iii**

Chapter 7
Host Plant Resistance

Multiple Choice Questions

1. **Who known as the father of Host Plant Resistance?**
   (a) B K Stain (b) B S Paul
   (c) R H Painter (d) T B Hans

2. **The inability of the insects to attack a non-host plant is termed as__________and such plant is not considered a host plant of the insect.**
   (a) Immunity (b) Resistant
   (c) Tolerant (d) Antifeedant

3. **First record of host plant resistance from New York, in 1782, from wheat variety "Underhill" against**
   (a) Hessain fly (b) Aphid
   (c) Termite (d) Cutworm

4. **Apple variety "Winter Majetin" in England found resistant against woolly aphid**
   (a) Mayetiola destructor (b) *Eriosoma lanigerum*
   (c) *Aphis gosspy* (d) *Myzus persicae*

5. **Procedure of host plant selection by insects**
   (a) Host habitat finding - Host recognition - Host finding - Host acceptance - Host suitability
   (b) Host habitat finding - Host finding - Host acceptance - Host recognition - Host suitability

(c) Host habitat finding - Host finding - Host recognition - Host suitability - Host acceptance

(d) Host habitat finding - Host finding - Host recognition - Host acceptance - Host suitability

6. **Insect herbivores are classified as Monophagous having**

(a) Single host (b) Restricted hosts

(c) Few hosts (d) Many hosts

7. **Insect herbivores are classified as Stenophagous having**

(a) Single host (b) Restricted hosts

(c) Few hosts (d) Many hosts

8. **Insect herbivores are classified as Oligophagous having**

(a) Single host (b) Restricted hosts

(c) Few hosts (d) Many hosts

9. **Insect herbivores are classified as Polyphagous having**

(a) Single host (b) Restricted hosts

(c) Few hosts (d) Many hosts

10. **Horizontal resistance is governed by**

(a) Single gene (b) Few genes

(c) Many genes (d) All of these

11. **Host plant selection is a process of chain reaction between**

(a) Plant response - Insect response (b) Plant stimuli - Insect response

(c) Insect stimuli - Plant response (d) Plant stimuli - Insect stimuli

12. **Resistance is heritable and controlled by**

(a) One gene (b) More major genes

(c) Few minor genes (d) Both (a) & (b)

13. **Resistance is relative and can be measured only by comparing with a susceptible cultivar of the**

(a) Different species (b) Same species

(c) Same genus (d) Same family

14. **Resistance is variable and can be modified by__________components of the environment.**

(a) Abiotic (b) Biotic

(c) Both (a) & (b) (d) None

15. Who classified the degree of resistance according to intensity?

(*a*) Stain (*b*) Paul
(*c*) Hans (*d*) Painter

16. Population capable of damaging and surviving on plants previously known to be resistant to other populations of the same species

(*a*) Biotype (*b*) Biospecies
(*c*) Biotic agents (*d*) None of these

17. The occurrence of asparagines in minute quantities was considered to be the primary cause of resistance to brown plant hopper in rice variety

(*a*) GEB 24 (*b*) IR 26
(*c*) TN 1 (*d*) Mudgo

18. From the Host Plant Resistance point of view Immunity, High Resistance, Low Resistance, Susceptibility and High Susceptibility are the

(*a*) Biotype (*b*) Intensity of resistance
(*c*) Plant response (*d*) Degree of intensity

19. Host may pass through the most susceptible stage quickly known as

(*a*) Host Evasion (*b*) Susceptibility
(*c*) High susceptibility (*d*) Low susceptibility

20. Heritable qualities of damage having a prior continuous, co-evolutionary history is referred as

(*a*) Sympatric resistance (*b*) Allopatric resistance
(*c*) Intrinsic resistance (*d*) Extrinsic resistance

21. Heritable qualities of damage having no prior continuous, co-evolutionary history is referred as

(*a*) Sympatric resistance (*b*) Allopatric resistance
(*c*) Intrinsic resistance (*d*) Extrinsic resistance

22. A trophic level resistance, plant alone produces defense through physical or chemicals means or both is called

(*a*) Sympatric resistance (*b*) Allopatric resistance
(*c*) Intrinsic resistance (*d*) Extrinsic resistance

23. A plant may pass through the most susceptible stage quickly or at time when insect number is less is called

(*a*) Host evasion (*b*) Induced resistance
(*c*) Escape (*d*) All of these

24. A trophic level resistance, where natural enemies of the pest are involved and benefit the host plant, is called

(a) Sympatric resistance (b) Allopatric resistance

(c) Intrinsic resistance (d) Extrinsic resistance

25. Which of the following is an allomone?

(a) Excitants (b) Attractants

(c) Repellents (d) All of these

26. A type of apparent host plant resistance in which a particular plant condition or environmental state makes a plant more resistant to pests than under other circumstances

(a) Acquired resistance (b) Permanent resistance

(c) Horizontal resistance (d) Induced resistance

27. When resistance is controlled by a single gene, it is called

(a) Monogenic resistance (b) Oligogenic resistance

(c) Polygenic resistance (d) None of these

28. When resistance is governed by a few genes, it is referred as

(a) Monogenic resistance (b) Oligogenic resistance

(c) Polygenic resistance (d) None of these

29. When resistance is governed by many genes, it is referred as

(a) Monogenic resistance (b) Oligogenic resistance

(c) Polygenic resistance (d) None of these

30. The resistance is controlled by one or few major genes called as

(a) Major gene resistance (b) Minor gene resistance

(c) Both (a) & (b) (d) None of these

31. When resistance is controlled by a number of minor genes, each contributing a small effect, it is called as

(a) Major gene resistance (b) Minor gene resistance

(c) Both (a) & (b) (d) None of these

32. The resistance attributed to morphological and anatomical factors referred as

(a) Alopatric resistance (b) Sympatric resistance

(c) Parapatric resistance (d) Phenetic resistance

33. The resistance mechanism employed by host plant to deter or to reduce colonization by insects is known as

(a) Antixenosis or non-preference (b) Antibiosis

(c) Tolerance (d) Susceptibility

34. Increase in resistance temporarily as result of some change conditions of plant or environment

(a) Host evasion
(b) Induced resistance
(c) Escape
(d) All of these

35. The earliest documented case of insect resistance is against

(a) Wheat stem sawfly
(b) Hessian fly
(c) Sorghum shoot fly
(d) Screw worm fly

36. Which of the following is a kairomone?

(a) Excitants
(b) Attractants
(c) Both (a) & (b)
(d) Repellents

37. The adverse effect of the host plant on the biology of the insects and their progeny, is known as

(a) Antixenosis or non-preference
(b) Antibiosis
(c) Tolerance
(d) Susceptibility

38. The ability of the host plant to withstand an insect population sufficient to damage severely the susceptible plants are known as

(a) Antixenosis or non-preference
(b) Antibiosis
(c) Tolerance
(d) Susceptibility

39. Phenetic resistance term coined by

(a) Kogan, 1994
(b) Paul, 1998
(c) Hans, 1985
(d) Painter, 1951

40. Solid stem of wheat resistant against

(a) Termite
(b) Cutworm
(c) Aphid
(d) Wheat stem sawfly

41. Physical resistance of sorghum showed against

(a) Stem borer
(b) Aphid
(c) Cutworm
(d) Sorghum shoot fly

42. Okra varieties had more and larger hairs on midrib and lamina of leaves showed resistance against

(a) Aphid
(b) Fruit borer
(c) Red cotton Bug
(d) Jassid

43. The occurrence of asparagines in minute quantities in rice variety Mudgo is considered resistant against

(a) Stem Borer
(b) Leaf Folder
(c) Gundhi Bug
(d) BPH

44. Non-nutritional chemical substances produced by one species and affects another species is known as

(a) Bio-chemicals
(b) Allelochemicals
(c) Toxic chemicals
(d) Non chemical

45. Allelochemicals which are advantageous to the producing organisms are known as

(a) Allomones
(b) Apneumone
(c) Kairomones
(d) None of these

46. Allelochemicals give adaptive advantageous to the receiving organisms are known as

(a) Allomones
(b) Apneumone
(c) Kairomones
(d) None of these

47. The major allelochemical imparting resistance to several insect pests in cotton is

(a) Solanin
(b) Tomatin
(c) Gossypol
(d) Capsicin

48. Cucurbitacins, chemically are a group of 20 Tetracyclic triterpenoids isolated from family

(a) Cucurbitaceae
(b) Solanaceae
(c) Malvaceae
(d) Leguminoceae

49. Which one of the following is a strategy to overcome development of biotypes?

(a) Pyramiding of major genes
(b) Incorporation of horizontal resistance
(c) Breeding tolerant varieties that exert no selection pressure on the insect
(d) All of these

50. The first BPH resistant rice variety with bph1 gene released in 1973 was

(a) IR 36
(b) IR 24
(c) IR 26
(d) Swarna

51. Quantitative trait governed by a large number of genes, each with a small contribution to resistance is referred as

(a) Monogenic resistance
(b) Polygenic resistance
(c) Oligogenic resistance
(d) None of these

52. Kogan and Ortman in 1965 was proposed the term

(a) Antixenosis
(b) Antibiotic
(c) Tolerance
(d) Resistances

53. First transgenic plant is

(*a*) Soybean (*b*) Cotton

(*c*) Tobacco (*d*) Maize

54. Tobacco transgenic plant was produced against Tobacco Hornworm, *Manduca sexta* via a vector

(*a*) Virus (*b*) Fungus

(*c*) Nematode (*d*) Agrobacterium

55. First country commercialize transgenic plants

(*a*) Australia (*b*) Brazil

(*c*) China (*d*) Denmark

56. Vertical resistance always governed by

(*a*) Major genes (*b*) Minor genes

(*c*) Both (*a*) & (*b*) (*d*) None of these

57. Horizontal resistance always governed by

(*a*) Major genes (*b*) Minor genes

(*c*) Both (*a*) & (*b*) (*d*) None of these

58. How many numbers of biotypes are found in Rice Gall Midge, *Orseolia oryzae?*

(*a*) 06 (*b*) 05

(*c*) 04 (*d*) 09

59. How many numbers of biotypes are found in wheat Hessain fly, *Mayetiola destructor?*

(*a*) 13 (*b*) 14

(*c*) 12 (*d*) 10

60. Maximum number of insect biotypes is observed in

(*a*) Brown plant hopper (*b*) Gall midge

(*c*) Hessian fly (*d*) Pea aphid

61. The nutrition value of the plant and absence of toxic compound help in the process of

(*a*) Host finding (*b*) Host acceptance

(*c*) Host recognition (*d*) Host suitability

62. Which of the following is a type of Ecological Resistance?

(*a*) Host evasion (*b*) Induced resistance

(*c*) Escape (*d*) All of these

63. How many numbers of biotypes are found in Apple woolly aphid, *Eriosoma lanigerum*?

(*a*) 03 (*b*) 04

(*c*) 05 (*d*) 06

64. How many numbers of biotypes are found in BPH of rice, *Nilaparvata lugens*?

(*a*) 03 (*b*) 05

(*c*) 04 (*d*) 06

65. How many numbers of biotypes are found in GLH of rice, *Nephotettix virescens?*

(*a*) 03 (*b*) 04

(*c*) 02 (*d*) 05

66. How many numbers of biotypes are found in Pea aphid, *Acyrthosiphon pisum*?

(*a*) 06 (*b*) 09

(*c*) 10 (*d*) 08

67. How many numbers of biotypes are found in Corn leaf aphid, *Rhopalosiphum maidis*?

(*a*) 05 (*b*) 04

(*c*) 06 (*d*) 07

68. Vertical resistance or Specific resistance and controlled by one or few

(*a*) Minor genes (*b*) Major genes

(*c*) Specific genes (*d*) None of these

69. The type of resistance is effective against certain specific biotypes of the insect but not others

(*a*) Horizontal resistance (*b*) Specific resistance

(*c*) Vertical resistance (*d*) Both (*b*) & (*c*)

70. DIMBOA is considered to be resistance factor against European corn borer in

(*a*) First generation (*b*) Second generation

(*c*) Third generation (*d*) Forth generation

71. Horizontal resistance or Non-specific resistance, it is controlled by many

(*a*) Major genes (*b*) Specific genes

(*c*) Minor genes (*d*) None of these

72. MBOA (6 - Methoxy - 2 - Benzoxazolinone) appears to function as a

(*a*) Repellent and/or feeding deterrent (*b*) Attractant

(*c*) Antibiotic (*d*) All of these

73. Maximum number of insect biotypes is observed in the insect group

(a) Aphids (b) Whiteflies

(c) Midges (d) Hoppers

74. In injured tissue DIMBOA converts to

(a) MBOA (b) MOA

(c) BOA (d) DIMA

75. Token stimuli theory was proposed by

(a) Fraenkel (b) House

(c) Throsteinson (d) Painter

76. Vertical resistance is also called

(a) Specific resistance (b) Major gene resistance

(c) Both (a) & (b) (d) None of these

77. The Rice variety Mudgo resistant to BPH because

(a) Proper nutrition (b) Nutritional imbalance

(c) Presence of toxic (d) Physical resistance

78. In barley resistance to green bug, *Schizaphis graminum* is associated with

(a) Presence of benzyl alcohol in leaves (b) High silica contents in leaves

(c) High MBOA contents in leaves (d) High saponin in leaves

79. The basis of antibiosis in corn against *Heliothis zea* is

(a) Presence of allomone (b) Nutritional imbalance

(c) Presence of toxic (d) Physical resistance

80. The basis of antibiosis in corn against European corn borer is

(a) Presence of growth inhibitor (b) Presence of feeding deterrent

(c) Presence of toxic (d) Physical resistance

81. Theory of chemosensory bases of host selection was proposed by

(a) Dethier (b) Snodgrass

(c) Hagstrum (d) Painter

82. A substance released by a non living material which is favourable to a receiving organism but detrimental to an organism of another species that is found in or on the non living material is known as

(a) Kairomone (b) Pheromone

(c) Synomone (d) Aneumone

83. The bio-type of BPH was first identified in Kerala on resistant rice variety

(a) IR - 8 (b) IR - 28

(c) IR - 18 (d) IR - 26

84. The relative amount of heritable qualities possessed by a plant that reduced the degree of damage done to the plant by a pest or pests is

(a) Host Plant Resistance (b) Tolerance

(c) Antibiosis (d) None of these

85. The type of resistance due to unfavourable environmental conditions for the inset is called as

(a) Pseudo resistance (b) Ecological resistance

(c) Induced resistance (d) All of these

86. Which one of the following is an advantage of host plant resistance?

(a) Specificity (b) Compatibility

(c) Decrease pesticide application (d) All of these

87. The type of resistance which is effective against all the known biotypes of the insects is known as

(a) Horizontal resistance (b) Non specific resistance

(c) Either a or b (d) None of these

88. Plant characteristics that drive insects away from a particular host is

(a) Antixenosis (b) Antibiosis

(c) Tolerance (d) Resistant

89. The ability of the plant to withstand an insect population sufficient to damage severely the susceptible plants is called

(a) Antixenosis (b) Antibiosis

(c) Tolerance (d) Resistant

90. Chemicals which prevent feeding of insects on a treated surface are called

(a) Antixenosis (b) Antibiosis

(c) Tolerance (d) Antifeedants

91. The chemicals which prevent insect damage to plants or animals by rendering them unattractive, unpalatable or offensive

(a) Antixenosis (b) Antibiosis

(c) Tolerance (d) Repellants

92. A non nutritional chemical, produced by an organism of one species and affect the growth, health, behavior or population biology of individuals of another species

(*a*) Allochemicals (*b*) Kairomone

(*c*) Synomone (*d*) Apneumone

93. A chemical released by an organism which benefits both the sender and the receiver is called

(*a*) Allochemicals (*b*) Kairomone

(*c*) Synomone (*d*) Apneumone

94. A multiple insect and disease resistant variety which is grown on more than 11 million ha in the world is

(*a*) IR 6 (*b*) IR 16

(*c*) IR 26 (*d*) IR 36

95. The occurrence of benzyl alcohol in minute quantities in barley was considered to be the primary cause of resistance to

(*a*) Green bug (*b*) Aphids

(*c*) Borer (*d*) None of these

96. Chemical that is released to the exterior by an organism and causes one or more specific reactions in a receiving organism of the same species is called

(*a*) Allochemicals (*b*) Kairomone

(*c*) Pheromone (*d*) Apneumone

97. Lethal Silk Factor of Host Plant Resistance is found against

(*a*) *Bombyx mori* (*b*) *Heliothis zea*

(*c*) *Helicoverpa armigera* (*d*) *Pectinophora gossypiella*

98. Which one of the following is limitation of host plant resistance?

(*a*) Time consuming (*b*) Biotype development

(*c*) Genetic limitations (*d*) All of these

Fill in the Blanks

1. "Dual Discrimination Theory" of Host Plant Resistance was proposed by__________.

2. "Dual Discrimination Theory" of Host Plant Resistance was based on insect's response to two types of stimuli, one is 'flavour stimuli' and another is__________.

3. "Token Stimuli Theory" of Host Plant Resistance was proposed by__________.

4. "Nutrient Imbalance Theory" of Host Plant Resistance was proposed by__________.

5. Those insects which are adapted to fit successfully into a variety of habitats or eat a variety of food are known as__________.

6. Those insects which have a narrow range of host plants and fed only on specialized plants are known as__________.

7. __________are the compounds which play an important role in co-evolutionary interaction of plant and insects.

8. __________represents a plant's relative ability to suffer a smaller loss of yield.

9. "Token Stimuli Theory" of Host Plant Resistance was based on__________.

10. __________denotes the plant characteristics that discourage insects for oviposition, food and shelter.

11. The compounds which are used in mediating plant - herbivore and herbivore - predator interaction as well as tritrophic interaction are called__________.

12. __________are the compounds released by organisms which produce detrimental reactions in receivers as well as emitters.

13. Chemicals which act as a feeding deterrent and growth inhibitor in insect pest but influence the host searching efficiency of predator and parasitoids are known as__________.

14. The gene produces lectin, is called__________.

Match the Following

1. **Match the following theories proposed by scientists**

(a) Dual Discrimination Theory	i) Kenedy and Booth (1951)
(b) Token Stimuli Theory	ii) House (1966)
(c) Nutrient Imbalance Theory	iii) Fraenkel (1959)
(d) Theory of Tolerance	iv) Painter (1951)

2. **Match the following type of insects with their host range**

(a) Monophagous	i) Sveral genera within a family
(b) Oligophagous	ii) Closely related species within a genera
(c) Polyphagous	iii) Restricted hosts
(d) Stenophagous	iv) Many hosts of different family

3. Match the following insects with their secondary metabolits

(a)	Pink boll worm	i)	Bombykol
(b)	American boll worm	ii)	Helilure
(c)	Mulberry silk worm	iii)	Gossyplure
(d)	Tobacco caterpillar	iv)	Litlure

4. Match the following hypothesis of Host Plant Resistance with their proposer

(a)	Resource concentration hypothesis	i)	Coley (1985)
(b)	Resource availability hypothesis	ii)	Root (1973)
(c)	The induced defense hypothesis	iii)	Price (1991)
(d)	The plant vigour hypothesis	iv)	Myers (1988)

Answers

Multiple Choice Questions

1	(c)	21	(b)	41	(d)	61	(d)
2	(a)	22	(c)	42	(d)	62	(d)
3	(a)	23	(d)	43	(d)	63	(a)
4	(b)	24	(d)	44	(b)	64	(b)
5	(d)	25	(c)	45	(a)	65	(a)
6	(a)	26	(d)	46	(c)	66	(b)
7	(b)	27	(a)	47	(c)	67	(a)
8	(c)	28	(b)	48	(a)	68	(d)
9	(d)	29	(c)	49	(d)	69	(d)
10	(c)	30	(a)	50	(c)	70	(a)
11	(b)	31	(b)	51	(b)	71	(c)
12	(d)	32	(d)	52	(a)	72	(a)
13	(b)	33	(a)	53	(c)	73	(a)
14	(c)	34	(b)	54	(d)	74	(a)
15	(d)	35	(b)	55	(c)	75	(a)
16	(a)	36	(c)	56	(a)	76	(c)
17	(d)	37	(b)	57	(b)	77	(b)
18	(b)	38	(c)	58	(a)	78	(a)
19	(a)	39	(a)	59	(b)	79	(a)
20	(a)	40	(d)	60	(c)	80	(b)

81	(a)	86	(d)	91	(d)	96	(c)
82	(d)	87	(c)	92	(a)	97	(b)
83	(d)	88	(a)	93	(c)	98	(d)
84	(a)	89	(c)	94	(d)		
85	(d)	90	(d)	95	(a)		

Fill in the Blanks

1. **Kenedy and Booth (1951)**
2. **Nutrient stimuli**
3. **Fraenkel (1959)**
4. **House (1966)**
5. **Eurytrophic/Euryphagic**
6. **Stenotrophic/Stenophagic**
7. **Allelochemicals**
8. **Tolerance**
9. **Secondary Biochemicals**
10. **Non-preference**
11. **Synemones**
12. **Antimones**
13. **Infochemicals**
14. **Lectin gene**

Match the Following

1. **a) - i, b) - iii, c) - ii, d) - iv**
2. **a) - ii, b) - i, c) - iv, d) - iii**
3. **a) - iii, b) - ii, c) - i, d) - iv**
4. **a) - ii, b) - i, c) - iv, d) - iii**

Chapter 8

Innovative Approaches in Pest Control

Multiple Choice Questions

1. **The utilization of chemicals that affect insect behaviour, growth or reproduction for suppression of insect population is known as**

   (*a*) Biological control (*b*) Botanical control

   (*c*) Biorational control (*d*) Viral control

2. **The term "Third Generation Pesticides" refers to**

   (*a*) Moulting hormones (*b*) Anti moulting hormones

   (*c*) Juvenile hormones (*d*) Anti Juvenile hormones

3. **The term "Fourth Generation Pesticides" refers to**

   (*a*) Moulting hormones (*b*) Anti moulting hormones

   (*c*) Juvenile hormones (*d*) Anti Juvenile hormones

4. **Which of the following pheromone is used for mating disruption technique?**

   (*a*) PB Rope - L (*b*) HL Rope

   (*c*) HP Rope (*d*) All of these

5. **The type of pheromone in red palm weevil, *Rhynchophorous ferrgineus* is**

   (*a*) Sex pheromone (*b*) Aggregation pheromone

   (*c*) Trail making pheromone (*d*) Alarm pheromone

6. A chemical which give adaptive advantage to the receiver but not to the emitter is

(*a*) Pheromone (*b*) Kairomone

(*c*) Synomone (*d*) Apneumone

7. Chitin is the most important structural component of the insect cuticle and is made up of N-acetyl-D-glucosamine unit jointed by

(*a*) β 2-4 linkage (*b*) β 1-4 linkage

(*c*) γ 1-4 linkage (*d*) α 1-4 linkage

8. The presence of chitin is restricted to

(*a*) Arthropods (*b*) Fungi and Nematodes

(*c*) Annelids (*d*) All of these

9. Chitin synthesis inhibitors are accidentally discovered during investigation of Dichlobenil is a/an

(*a*) Insecticide (*b*) Herbicide

(*c*) Fungicide (*d*) Nematicide

10. Diflubenzuron [1-(4-chlorophenyl)-3-(2,6-dipluorbenzoyl)urea] was commercialized under the name

(*a*) Dichlobenil (*b*) Biolep

(*c*) Dipel (*d*) Dimilin

11. Dimilin is effective against insect pest belongs to

(*a*) Coleoptera (*b*) Diptera

(*c*) Lepidoptera (*d*) All of these

12. The Benzoyl Phenyl Urea (BPU) analogues act as

(*a*) Stomach toxicants (*b*) Sterilants

(*c*) Ovicides (*d*) All of these

13. During the process of biosynthesis of chitin the BPU analogs, blocks the terminal polymerisation step catalyzed by the enzyme

(*a*) Polymerase (*b*) Catalase

(*c*) Chitin synthase (*d*) Anolase

14. Plumbagin is a naturally occurring__________present in the roots of *Plumbago capensis*, which inhibits ecdysis in several lepidopteran pests including *Helicoverpa* sp.

(*a*) Juvenile hormone (*b*) Chitin synthesis inhibitor

(*c*) Antifeedant (*d*) Mating disrupting agent

15. Buprofezin has been found effective against several

(*a*) Lepidopteran pests (*b*) Coleopteran pests
(*c*) Hemipteran pests (*d*) Orthopteran pests

16. Buprofezin act on insects as Chitin Synthesis Inhibitor (CSI) and inhibit the biosynthesis of

(*a*) Chitin (*b*) Prostaglandin
(*c*) Syntheses (*d*) Cuticulin

17. Old chitin is degenerated and new chitin is laid down again at each

(*a*) Stadium (*b*) Generation
(*c*) Moult (*d*) Morph

18. Degeneration of chitin is carried out by chitinolytic enzymes

(*a*) Chitinase (*b*) Chitobiase
(*c*) Syntheses (*d*) Both (*a*) & (*b*)

19. The chitinase gene is, however, being considered a favourable target for

(*a*) Molecular biology (*b*) Microbiology
(*c*) Genetic engineering (*d*) None of these

20. The only available sclerotization inhibitor is__________developed by Monsento.

(*a*) MON 0585 (*b*) MON 025
(*c*) MON 5805 (*d*) MON T 058

21. MON 0585 chemically is a di-tertiary butyl alcohol which is highly toxic to

(*a*) Dipteran pests (*b*) Lepidopteran pests
(*c*) Coleopteran pests (*d*) Orthopteran pests

22. DDC (3,4 - dihydroxyphenyl alanine decarboxylase) is an enzyme of

(*a*) Sclerotization (*b*) Moulting
(*c*) Ecdysis (*d*) Metamorphosis

23. A chemical which give adaptive advantage to the emitter but not to the receiver is

(*a*) Pheromone (*b*) Kairomone
(*c*) Synomone (*d*) Apneumone

24. When the larvae moults into pupa the concentration of Juvenile Hormone is

(*a*) Increased (*b*) Decreased
(*c*) Remains unchanged (*d*) None of these

25. Inhibitor of DDC enzyme, causing mortality in dipteran larvae at the time of moulting is

(*a*) MON 0585 (*b*) MON 025

(*c*) 5805 methyl DOPA (*d*) α - methyl DOPA

26. Chemical involved for accelerator of sclerotization process are

(*a*) Scleretizine (*b*) Cryomazine

(*c*) Ecdyson (*d*) Bursicon

27. Cryomazine is a substituted di-amino-triazine inserting an extra layer between the endocuticle and__________causing rod like puparia.

(*a*) Exocuticle (*b*) Procuticle

(*c*) Mesocuticle (*d*) Basement membrane

28. The life processes of insects are regulated mainly by__________and endocrine systems.

(*a*) Neural (*b*) Skeletal

(*c*) Water Vascular (*d*) None of these

29. Hormones involved in regulating the life processes of insects are

(*a*) Brain hormone (*b*) Moulting hormone

(*c*) Juvenile hormone (*d*) All of these

30. The brain hormone or prothoracicotropic hormone (PTTH) secreted by

(*a*) Neurosecretory cells (*b*) Prothoracic glands

(*c*) Corpora allata (*d*) Corpora cardiaca

31. Which of the following tool is/are used to measure the behavior of insect towards the odour source

(*a*) Wind tunnel (*b*) Y olfactometer

(*c*) Both (*a*) & (*b*) (*d*) None of these

32. In which of the following insect species males produce pheromone

(*a*) *Trichoplusia ni* (*b*) *Danaus glippus*

(*c*) *Ceratitis capitata* (*d*) All of these

33. Which of the following is a chemosterilant

(*a*) TEPA (*b*) METATEPA

(*c*) Thio - TEPA (*d*) All of these

34. PTTH activates the__________which secrete Moulting Hormone (MH) or Ecdysone

(*a*) Neurosecretory cells (*b*) Prothoracic glands

(*c*) Corpora allata (*d*) Corpora cardiaca

35. Juvenile hormone secreted by the

(a) Neurosecretory cells (b) Prothoracic glands

(c) Corpora allata (d) Corpora cardiaca

36. In presence of high titre of Juvenile Hormone, a larva moults into a larva, while at low titre of JH it moults into a

(a) Larva (b) Pupa

(c) Adult (d) None of these

37. In the absence of Juvenile Hormone, the pupae moult to become

(a) Larva (b) Pupa

(c) Adult (d) None of these

38. Tanning hormone or bursicon regulates the process of tanning and sclerotization of the

(a) Old cuticle (b) New cuticle

(c) Eye (d) Bristles

39. Brain hormones or neurohormones are produced by the

(a) Central Nervous System (b) Periferal Nervous System

(c) Viseral Nervous System (d) None of thesse

40. First insect neuropeptide proctolin was isolated from

(a) *Periplaneta americana* (b) *Helicoverpa armigera*

(c) *Musca domestica* (d) *Bombyx mori*

41. The brain hormones chemically either small protein molecules or

(a) Oligopeptides (b) Monosaccharide

(c) Polypeptides (d) Disaccharides

42. Removal of neuropeptides can lead to mortality of the affected insects because these exert critical control over__________functions of insects.

(a) Development (b) Reproductive

(c) Metabolic and Behavioural (d) All of these

43. In higher diptera juvenile hormone is found as

(a) JH III - bisepoxide (b) 4 - methyl JH I

(c) JH I (d) JH II

44. In insects juvenile hormone regulates

(a) Metamorphosis (b) Vitellogenesis

(c) Diapause and polymorphism (d) All of these

45. The juvenile hormones are terpenoidal compounds synthesized from__________via farnesyl pyrophosphate (FPP).

(*a*) Asyle CoA (*b*) Acetyl CoA

(*c*) Glucose (*d*) None of these

46. Which of the following is repellent to mosquitoes?

(*a*) DEET (*b*) Oil of citronella

(*c*) Dimethyl phthalate (*d*) All of these

47. Which of the following is a chitin synthesis inhibitor?

(*a*) Buprofezin (*b*) Teflubenzuron

(*c*) Flufenoxuron (*d*) All of these

48. Earlier commercial Juvenile Hormone Analogue (JHA) was synthesized and used for pest control are (where 2,4 - dienoates group are present)

(*a*) Methoprene (*b*) Hyloprene

(*c*) Bursicon (*d*) None of these

49. Isoprene groups have been substituted by linked phenoxy groups in newer JHA, like

(*a*) Fenoxycarb (*b*) Pyriproxyfen

(*c*) Hydroprene (*d*) Both (*a*) & (*b*)

50. Juvabione are Juvenile Hormone Analogue found from__________.

(*a*) Plant products (*b*) Animal

(*c*) Bacteria (*d*) Virus

51. JHA found as, methyl ester of todomatuic acid and its analogue, dehydrojuvabione are found from the plant Balsam fir tree,

(*a*) *Abies balsamea* (*b*) *Sorea robusta*

(*c*) *Terminalia arjuna* (*d*) *Butea monosperma*

52. Application of JHA to adult females act as

(*a*) Ovicidal (*b*) Precocine

(*c*) Moulting (*d*) None of these

53. Anti juvenile hormone agents or JH antagonists act in promotion of__________development in insects.

(*a*) Normal (*b*) Precocious

(*c*) Delay (*d*) No development

54. Moulting hormones are chemically__________compounds secreted by prothoracic glands.

(*a*) Starch (*b*) Protein

(*c*) Steroidal (*d*) Fat

55. Tebufenozide is a moulting hormone analogue used as an insecticide against Larval stages of

(*a*) Lepidopteran insects (*b*) Dipteran insects
(*c*) Coleopteran insects (*d*) Hymenopteran insects

56. The intraspecific semiochemicals are known as

(*a*) Pheromone (*b*) Kairomones
(*c*) Synomone (*d*) Allomones

57. Interspecific semiochemicals, benefits receiver are known as

(*a*) Pheromone (*b*) Kairomones
(*c*) Synomone (*d*) Allomones

58. Allomones are semiochemicals which benefits

(*a*) Receivers (*b*) Emitters
(*c*) Both (*a*) & (*b*) (*d*) None of these

59. Interspecific semiochemicals which benefits both receiver and emitters are known as

(*a*) Pheromone (*b*) Kairomones
(*c*) Synomone (*d*) Allomones

60. The term pheromone was proposed by__________in 1959.

(*a*) Painter (*b*) Karlson and Luscher
(*c*) C. Linnaeus (*d*) Whittaker

61. A chemical not found in an insect has a pheromone like action, it is often referred to as a

(*a*) Parapheromone (*b*) Kairomones
(*c*) Synomone (*d*) Allomones

62. Chemicals involved in interspecific communication were given the name of

(*a*) Allelochemicals (*b*) Semiochemicals
(*c*) Pheromone (*d*) Synomone

63. The term Allelochemicals was proposed by__________in 1970.

(*a*) Painter (*b*) Karlson and Luscher
(*c*) C. Linnaeus (*d*) Whittaker

64. A substance emitted by a non-living material which evokes a behavioural or physiological adaptively favourable to a receiving organism but detrimental to an organism of another species is known as

(*a*) Synomone (*b*) Apneumone
(*c*) Pheromone (*d*) Kairomone

65. Aggregation pheromone is a substance produced by one or both sexes and bringing both sexes together for

(*a*) Feeding (*b*) Reproduction

(*c*) Flying (*d*) Both (*a*) & (*b*)

66. A substance produced by an insect to repel and disperse other insects in the area are known as__________, it is usually released by an individual when it is attacked.

(*a*) Alarm pheromone (*b*) Para pheromone

(*c*) Aggregation pheromone (*d*) Sex pheromone

67. Most of the parapheromones in use today as male attractants of *Bractocera* are

(*a*) Trimedlure (*b*) Methyl eugenol

(*c*) Cue-lure (*d*) All of these

68. Anti juvenile hormone agents are

(*a*) Precocene I and II (*b*) Precocene VI and VII

(*c*) Precocene V and VI (*d*) Precocene IV and III

69. Nereistoxin is a dithiolane compound found naturally in the marine annelid, *Lumbriconereis heteropoda* it act as a

(*a*) Non-depolarizing neuromuscular blocking agent

(*b*) Depolarizing neuromuscular blocking agent

(*c*) Non neuromuscular blocking agent

(*d*) Fumigants

70. Cartap is a pro-insecticide, in insect body which is rapidly converted to

(*a*) Ethyl alcohol (*b*) Nereistoxin

(*c*) Non toxic metabolites (*d*) Hydrochloride

71. Avermectins now used as insecticide, were discovered from a culture of

(*a*) *Streptomyces avermitilis* (*b*) *L. heteropoda*

(*c*) *Bractocera* spp. (*d*) *Saccharopolyspora spinosa*

72. Chemically avermectins are

(*a*) Macrocyclic lactones (*b*) Tetrano triterpinoids

(*c*) Chlorinated hydrocarbon (*d*) None of these

73. Successfully commercialized avermectins are

(*a*) Ivermectin (*b*) Abamectin (Avermectin B_1)

(*c*) Ibamectin (*d*) Both (*a*) & (*b*)

74. Ivermectin has been successfully used to control

(*a*) Plant parasites (*b*) Animal parasites

(*c*) Bacterial diseases (*d*) Fungal diseases

75. Abamectin is rapidly absorbed by plant tissue, so it does not persist as contact poison, it used as a

(*a*) Fumigants (*b*) Systemic poison

(*c*) Antifeedant (*d*) Mating disrupting

76. To control insects, nematode, mite etc. Abamectin is

(*a*) Highly selective (*b*) Broad spectrum

(*c*) Non selective (*d*) Relatively non toxic

77. Spinosad (Spinosyns) are naturally derived from *Saccharopolyspora spinosa* is a/an

(*a*) Bacteria (*b*) Actinomycetes

(*c*) Fungi (*d*) Lichen

78. Spinosad, a mixture of spinosyn A and spinosyn B is the lead compound and it has shown

(*a*) Contact activity (*b*) Stomach activity

(*c*) Fumigant activity (*d*) Both (*a*) & (*b*)

79. The genetic control methods utilize to insect pests for their own destruction are also known as

(*a*) Auto confusing methods (*b*) Autocidal methods

(*c*) Auto immune method (*d*) All of these

80. In which of the following insect species male produce pheromone

(*a*) Red palm weevil (*b*) Banana corm weevil

(*c*) Cotton boll weevil (*d*) All of these

81. Which of the following is used for mating disruption of pink bollworm of cotton?

(*a*) PB Rope - L (*b*) PG Rope - L

(*c*) HL Rope (*d*) HP Rope - 536

82. The genetic control methods of insect pest control consists of

(*a*) Sterile Insect Technique

(*b*) Incompatible Insect Technique

(*c*) Use of Hybrid Sterility

(*d*) All of these

83. The use of genetic method to control insect pests was started with *Cochliomya homnivorax* eradication programme is a

(*a*) Screw worm fly (*b*) Fruit fly

(*c*) House fly (*d*) Blow worm fly

84. The use of genetic method to control insect pests was started with *Cochliomya homnivorax* eradication programme by

(*a*) AS Atwal in 1939 (*b*) EF Knipling in 1940

(*c*) RH Painter in 1951 (*d*) AD Imms 1952

85. Which of the followings act as Chemosterilants, used for control of insect pests?

(*a*) TEPA (*b*) Hempa

(*c*) Busulfan (*d*) All of these

86. All effective chemosterilants have not found favour in IPM programmes because in man, these chemicals display strong__________action.

(*a*) Carcinogenic (*b*) Mutagenic

(*c*) Teratogenic (*d*) Both (*a*) & (*b*)

87. Repellents are chemicals that prevent insect damage to plants by rendering them

(*a*) Unattractive (*b*) Unpalatable

(*c*) Offensive (*d*) All of these

88. The main groups of Insect Growth Regulators (IGRs) being used on a commercial scale are

(*a*) Juvenile Hormone Analogues (*b*) Anti Juvenile Hormones

(*c*) Chitin Synthesis Inhibitor (*d*) Both (*a*) & (*c*)

89. The biorational insecticides rapidly lose their efficacy because they are

(*a*) Photostable (*b*) Photodegradable

(*c*) Thermostable (*d*) None of these

90. Pink bollworm *Pectinophora gossypiella* a key pest of cotton can be successfully managed by using controlled release formulations of its female sex pheromone, known as

(*a*) Pectilure (*b*) Gossyplure

(*c*) Culure (*d*) None of these

91. __________is a Chitin Synthesis Inhibitor (CSI), besides this it also found to act as an inhibitor of Ecdysteroid biosynthesis.

(*a*) BPU (*b*) Buprofezin

(*c*) Plumbagin (*d*) All of these

92. Which of these is not considered as an insect growth regulator?
(*a*) Synthetic pyrethroid (*b*) Chitin inhibitor
(*c*) Juvenile hormone analogue (*d*) Ecdysteroid

93. Generally Pheromone trap attracts
(*a*) Females (*b*) Males
(*c*) Workers (*d*) None of these

94. The name of EF Knipling is associated with
(*a*) IPM (*b*) Sterile insect technique
(*c*) Pheromone (*d*) Apiculture

95. The screw wormfly (*Cochliomyia hominivorax*) was kept under control through
(*a*) Synthetic pyrithroids (*b*) DDT
(*c*) Sterile insect technique (*d*) IPM

96. The hypothetical ratio of male and female subjected to SIRM
(*a*) 1 : 1 (*b*) 1 : 2
(*c*) 2 : 1 (*d*) 1 : 3

97. The role of Juvenile Hormone in growth of insect was first reported by
(*a*) Williams (*b*) Steiner
(*c*) Wigglesworth (*d*) None of these

98. "Paper factor" is associated with
(*a*) Juvenile Hormone (*b*) Pheromone
(*c*) Hormone (*d*) DDT

99. Trimedlure is associated with
(*a*) Paddy gallfly (*b*) Oriental fruitfly
(*c*) Melon fruitfly (*d*) Mediterranean fruitfly

100. R H Painter is associated with
(*a*) SIRM (*b*) Bilogical control
(*c*) Host Plant Resistance (*d*) None of these

101. Methyl eugenol is used to attract the males of
(*a*) Melon fruitfly (*b*) Oriental fruitfly
(*c*) Paddy gallfly (*d*) Mediterranean fruitfly

102. Example of insect hormone is
(*a*) Ecdysone (*b*) Chitin
(*c*) Lindane (*d*) Allethrin

103. Autocidal control technique of insect control was proposed by

(*a*) Lefroy (*b*) Pruthi

(*c*) Frisch (*d*) Knipling

104. Example of repellent used against termites is

(*a*) DEET (*b*) Citronella oil

(*c*) Penta chlorophenol (*d*) Neem oil

105. A substance which make the plant distasteful to insect feeding is called

(*a*) Antifeedant (*b*) Repellent

(*c*) Attractant (*d*) Pheromone

106. Name of the precursor of ecdysone synthesis is

(*a*) Ergosterol (*b*) Methionine

(*c*) Cholesterol (*d*) Arachidic acid

107. Non preference to host plant is also known as

(*a*) Antibiosis (*b*) Antixenosis

(*c*) Tolerance (*d*) None of these

108. Which crop has maximum area under transgenic crop in the world?

(*a*) Sugarcane (*b*) Cotton

(*c*) Tobacco (*d*) Soybean

109. The gene Bar, Barstar, Barnase are used for developing form GM

(*a*) Soybean (*b*) Cotton

(*c*) Mustard (*d*) Maize

110. Genetic Engineering approval committee comes under the ministry of

(*a*) Environment and Forests (*b*) Agriculture

(*c*) Science and Technology (*d*) Rural Development

111. Protein hydrolyzates used as an attractant for the control of

(*a*) White fly (*b*) Gall fly

(*c*) House fly (*d*) Fruit fly

112. Which of the following is an example of chitin synthesis inhibitor?

(*a*) Dimilin (*b*) Biolep

(*c*) Biomass (*d*) Spiromycin

113. Sterile insect control method is most effective against

(*a*) Honey bee (*b*) Housefly

(*c*) Termite (*d*) None of these

114. Pheromone is also known as

(*a*) Ectohormone (*b*) Hormone

(*c*) Kairomone (*d*) None of these

115. According to Insecticide Act, 1968, Pheromones used for the following purpose requires registration

(*a*) Monitoring (*b*) Mass trapping

(*c*) Mating disruption (*d*) Both (*a*) & (*b*)

116. Some Lepidopteran pests produce pheromones by Androconail glands present in their

(*a*) Legs (*b*) Antennae

(*c*) Wings (*d*) Scales

117. The aggregation pheromone produced by the male cotton boll weevil, *Anthonomus grandis* is

(*a*) Cuelure (*b*) Grandlure

(*c*) Ferrugineol (*d*) Cosmolure

118. The aggregation pheromone produced by red palm weevil, *Rhynchophorous ferrugineus* is

(*a*) Cuelure (*b*) Grandlure

(*c*) Ferrugineol (*d*) Cosmolure

119. The aggregation pheromone produced by banana corm weevil, *Cosmopolites sordidus* is

(*a*) Cuelure (*b*) Grandlure

(*c*) Ferrugineol (*d*) Cosmolure

120. Which of the following is juvenile hormone analog?

(*a*) Hydroprene (*b*) Juvabione

(*c*) Methoprene (*d*) All of these

121. Pheromones used for the following purpose requires no registration

(*a*) Monitoring (*b*) Mass trapping

(*c*) Mating disruption (*d*) Both (*a*) & (*b*)

122. Which of the following is a chitin synthesis inhibitor?

(*a*) Diflubenzuron (*b*) Chlorfluazuron

(*c*) Teflubenzuron (*d*) All the above

Fill in the Blanks

1. The utilization of chemicals that affect insect behaviour, growth or reproduction for suppression of insect population is known as__________.

2. In each moult old chitin is degenerated and__________is laid down again.

3. The chitinase gene is being considered as a favorable target for__________.

4. Juvenile hormone is secreted by the__________.

5. Prothoracicotropic hormone activates the__________gland, which secrete moulting hormone.

Match the Following

1. Match the following insect growth regulator with their trade name

(a) Buprofezin	i) Atabron
(b) Chlorfluzuron	ii) Dimilin
(c) Diafenthiuron	iii) Applaud
(d) Diflubenzuron	iv) Polo

2. Match the following insect growth regulator with their trade name

(a) Flufenoxuron	i) Rimon
(b) Lufenuron	ii) Match
(c) Novaluron	iii) Cascade
(d) Teflubenzuron	iv) Nomolt

3. Match the alarm pheromone with respective insect

(a) Formic acid	i) Aphids
(b) Terpenes	ii) Ants
(c) Monoterpene hydrocarbons	iii) Hemipterans
(d) Aldehydes	iv) Termite soldiers

4. Match the aggregation pheromone with respective insect

(a) Frontalin	i) *Tribolium*
(b) Ipsenol	ii) *Ips*
(c) Periplanone	iii) *Periplaneta*
(d) Dimethyldecanol	iv) *Dendroctonus*

5. **Match the following arthropods with their repellents**

(a) Termites	i) Pentchlorophenol
(b) Mosquitoes	ii) Benzyl benzoate
(c) Cloth moths	iii) Napthalene
(d) Mites	iv) Citronellol

6. **Match the following insects with their repellents**

(a) Crawling insects	i) Bordeaux mixure
(b) Phytophagous insects	ii) Creosote
(c) Wood feeders	iii) Dimethyl phthalate
(d) Blood sucking insects	iv) Pentachlorophenol

7. **Match the following insects with their attractants**

(a) Cabbage butterfly	i) Sinigrin
(b) House fly	ii) Molasses
(c) Bark beetle	iii) Terpenes
(d) Japanese beetle	iv) Geraniol

8. **Match the following insects with their attractants**

(a) Onion fly	i) Cuelure
(b) Oriental fruit fly	ii) Trimedlure
(c) Melon fruit fly	iii) Propylmercaptan
(d) Mediterranean fruit fly	iv) Methyl eugenol

Answers

Multiple Choice Questions

1	(c)	21	(a)	41	(a)	61	(a)
2	(c)	22	(a)	42	(d)	62	(a)
3	(d)	23	(a)	43	(a)	63	(d)
4	(d)	24	(b)	44	(d)	64	(b)
5	(b)	25	(d)	45	(b)	65	(d)
6	(b)	26	(b)	46	(d)	66	(a)
7	(b)	27	(a)	47	(d)	67	(d)
8	(d)	28	(a)	48	(a)	68	(a)
9	(b)	29	(d)	49	(d)	69	(a)
10	(d)	30	(a)	50	(a)	70	(b)
11	(d)	31	(c)	51	(a)	71	(a)
12	(d)	32	(d)	52	(a)	72	(a)
13	(c)	33	(d)	53	(b)	73	(d)
14	(b)	34	(b)	54	(c)	74	(b)
15	(c)	35	(c)	55	(a)	75	(b)
16	(b)	36	(b)	56	(a)	76	(a)
17	(c)	37	(c)	57	(b)	77	(b)
18	(d)	38	(b)	58	(b)	78	(d)
19	(c)	39	(a)	59	(c)	79	(b)
20	(a)	40	(a)	60	(b)	80	(d)

81	(a)	92	(a)	103	(d)	114	(a)
82	(d)	93	(b)	104	(c)	115	(c)
83	(a)	94	(b)	105	(a)	116	(c)
84	(b)	95	(c)	106	(c)	117	(b)
85	(d)	96	(a)	107	(b)	118	(c)
86	(d)	97	(a)	108	(d)	119	(d)
87	(d)	98	(a)	109	(c)	120	(d)
88	(d)	99	(d)	110	(a)	121	(d)
89	(b)	100	(c)	111	(d)	122	(d)
90	(b)	101	(b)	112	(a)		
91	(c)	102	(a)	113	(d)		

Fill in the Blanks

1. Biorational control
2. New chitin
3. Genetic engineering
4. Corpora allata
5. Prothoracic

Match the Following

1. a) - iii, b) - i, c) - iv, d) - ii
2. a) - iii, b) - ii, c) - i, d) - iv
3. a) - ii, b) - i, c) - iv, d) - iii
4. a) - iv, b) - ii, c) - iii, d) - i
5. a) - i, b) - iv, c) - iii, d) - ii
6. a) - ii, b) - i, c) - iv, d) - iii
7. a) - i, b) - ii, c) - iii, d) - iv
8. a) - iii, b) - iv, c) - i, d) - ii

Chapter 9

Integrated Pest Management

Multiple Choice Questions

1. IPM stands for

(*a*) Insect Pest Management
(*b*) Integrated Pest Monitoring
(*c*) Internal Pest Management
(*d*) Integrated Pest Management

2. Insect pest control is essential because

(*a*) They transmit plant disease
(*b*) They feed on valuable crops
(*c*) Insect creates nuisance
(*d*) All of these

3. Which of the following insect of mango has quarantine importance.

(*a*) Stone or nut weevil
(*b*) Fruit fly
(*c*) Mealy bug
(*d*) Leafhopper

4. Use of resistant varieties in the IPM is an example of

(*a*) Biological control
(*b*) Legal control
(*c*) Cultural control
(*d*) Physical control

5. ETL stands for

(*a*) Economic Threshold Level
(*b*) Economic Threshing Level
(*c*) Economic Thrust Level
(*d*) None of these

6. Pest population at Economic Injury Level is__________than the Economic Threshold Level.

(*a*) Higher
(*b*) Lower
(*c*) Equal
(*d*) Not detected

7. Any living organism which causes directly or indirectly damage to human kind is known as

(*a*) Insect (*b*) Pest

(*c*) Rodents (*d*) Disease

8. An important major pest species in the complex of pests attacking a crop is known as

(*a*) Key pests (*b*) Regular pests

(*c*) Sporadic pests (*d*) Endemic pest

9. These pests are generally found in abundance during a crop season is known as

(*a*) Regular pests (*b*) Key pests

(*c*) Sporadic pests (*d*) Endemic pest

10. These are those pests which appear occasionally in certain years and cause little damage is known as

(*a*) Sporadic pests (*b*) Key pests

(*c*) Regular pests (*d*) Endemic pest

11. The pest which are restricted to a particular locality is known as

(*a*) Endemic pest (*b*) Key pests

(*c*) Regular pests (*d*) Sporadic pests

12. These pest normally cause negligible damage but may become highly destructive resulting from some disturbance in the environment and the consequent increase in their number is known as

(*a*) Potential pest (*b*) Occasional pest

(*c*) Persistent pest (*d*) Perennial pests

13. These pests which reaches significant level only occasionally is known as

(*a*) Potential pest (*b*) Occasional pest

(*c*) Persistent pest (*d*) Perennial pest

14. These pest are present throughout the crop growth is known as

(*a*) Potential pest (*b*) Occasional pest

(*c*) Persistent pest (*d*) Perennial pests

15. The pest whose economic injury level are only slightly about the general equilibrium position and intervention is necessary at every stage when the population reaches upward level is known as

(*a*) Potential pest (*b*) Occasional pest

(*c*) Persistent pest (*d*) Perennial pests

16. YSB of rice, American Boll Worm of cotton, DBM on cabbage, fall under the category

(*a*) Endemic pest (*b*) Key pests

(*c*) Regular pests (*d*) Sporadic pests

17. The level of injury where damage can be measured is called as

(*a*) Damage Boundary (*b*) General Equilibrium

(*c*) Economic Threshold (*d*) Economic Injury

18. The concept of Economic Injury Level was first proposed by

(*a*) Stern *et al.* (1959) (*b*) Paul *et al.* (1958)

(*c*) Frank *et al.* (1957) (*d*) Stain *et al* (1956)

19. For their pioneer work on development and implementation of IPM concept, awarded World Food Prize during 1997

(*a*) Roy F Smith and Perry L Adkisson (*b*) Paul D S and Roy T

(*c*) Frank E L and Lal B T (*d*) Stain A L and Das A S

20. In coconut, the rhinoceros beetle can be picked out of the holes with the help of crooked hooks made of iron this is a type of

(*a*) Cultural control (*b*) Physical control

(*c*) Chemical control (*d*) Mechanical control

21. National Plant Quarantine Station (NPQS) is located at

(*a*) New Delhi (*b*) Hyderabad

(*c*) Mumbai (*d*) Amritsar

22. Digging of trenches around the fields is useful for protecting the pest

(*a*) Locust (*b*) Grasshoppers

(*c*) Red hairy caterpillar (*d*) All of these

23. Advantage of using resistance variety

(*a*) Specificity (*b*) Easily adoptable

(*c*) Conflicting traits (*d*) Both (*a*) & (*b*)

24. Hand picking and use of various traps are example of

(*a*) Mechanical control (*b*) Physical control

(*c*) Cultural control (*d*) All of these

25. Pest outbreaks tend to occur when

(*a*) Crops are planted in monoculture

(*b*) Natural enemies population are low

(*c*) Favourable climatic condition

(*d*) All of these

26. Which control strategy is likely to have the greatest impact on non target organism?

(a) Chemical control (b) Biological control

(c) Cultural control (d) Physical control

27. National Biodiversity Authority (NBA) is located at

(a) Hyderabad (b) New Delhi

(c) Mumbai (d) Chennai

28. Insecticide management is required to

(a) Reduce the load of chemical insecticide

(b) Restrict the use of chemical insecticide

(c) No use of chemical insecticide

(d) Use of biological insecticide

29. Resurgence of cotton whitefly occurred due to the use of

(a) Synthetic pyrithroid (b) Carbofuran

(c) Quinalphos (d) Phorate

30. National Plant Protection Training institute (NPPTI) is located at

(a) Hyderabad (b) New Delhi

(c) Mumbai (d) Chennai

31. The average population density of an insect population over a long period of time, unaffected by the temporary interventions of pest control is known as

(a) General Equilibrium Position (b) Damage Boundary

(c) Economic Injury (d) Economic Threshold

32. The lowest degree of pest population that does economically significant damage is known as

(a) Economic Injury Level (b) General Equilibrium

(c) Damage Boundary (d) Economic Threshold

33. The density of pest population at which control measure should be applied to prevent the pest from reaching the economic injury level is known as

(a) General Equilibrium Position (b) Damage Boundary

(c) Economic Injury (d) Economic Threshold Level

34. Domestic quarantine were enforced in India to prevent the spread of

(a) Apple woolly aphid (b) Pink bollworm

(c) Brown plant hopper (d) San Jose Scale

35. Pest population that produces incremental damage equal to the cost of preventing damage is

(*a*) Economic Injury Level (*b*) Economic Threshold Level
(*c*) General Equilibrium Position (*d*) None of these

36. Alternate drying and wetting of paddy fields to reduce BPH population is a method of

(*a*) Physical control (*b*) Mechanical Control
(*c*) Biological Control (*d*) Cultural control

37. The insect which entered India before the enforcement of quarantine measures is

(*a*) Cottony cushion scale (*b*) Cotton whitefly
(*c*) Guava mealy bug (*d*) Sugarcane pyrilla

38. In mango, sticky bands around tree trunks provide protection against

(*a*) Hoppers (*b*) Bark beetle
(*c*) Mango mealy bugs (*d*) All of these

39. Use of yellow sticky trap coated with castor oil to trap white fly is a method of

(*a*) Physical control (*b*) Mechanical control
(*c*) Biological control (*d*) Cultural control

40. Use of sticky traps where stick glue is pasted is a method of

(*a*) Physical control (*b*) Mechanical control
(*c*) Biological control (*d*) Cultural control

41. Use of sticky traps where stick glue is pasted to trap

(*a*) Aphids (*b*) Beetle
(*c*) Cutworm (*d*) Damsel fly

42. Pest of potato which has gained entry into India

(*a*) Potato Tuber Moth (*b*) Cutworm
(*c*) Potato aphid (*d*) None of these

43. In chick pea early sowing is done to protect the crop from which of the following pest

(*a*) *Helicoverpa armigera* (*b*) Aphids
(*c*) Cut worms (*d*) None of these

44. Sterile Insect Technique (SIT), developed by E F Knipling and R C Bushland was first used to eliminate

(*a*) Screwworms (*b*) Mealworms
(*c*) Mosquitoes (*d*) Roundworms

45. Scientific name of screwworm fly is

(*a*) *Helicoverpa armigera* (*b*) *Cotecia rufomaclata*

(*c*) *Cochliomyia hominivorax* (*d*) *Ceratitis capitata*

46. The National Centre for Integrated Pest Management (NCIPM) is located at

(*a*) Hyderabad (*b*) Chandigarh

(*c*) New Delhi (*d*) Pune

47. Which one of the following viral diseases is recently introduced in India from a neighbouring country?

(*a*) Papaya leaf curl (*b*) Cotton leaf curl

(*c*) Chilli leaf curl (*d*) Tomato leaf curl

48. *Leusinodes orbonalis* is a pest of

(*a*) Paddy (*b*) Cotton

(*c*) Brinjal (*d*) Tea

49. Cause of pest outbreak

(*a*) Large scale monoculture (*b*) Favourable weather condition

(*c*) Both (*a*) & (*b*) (*d*) Can't be said

50. Natural enemies are used in pest control is the example of

(*a*) Chemical control (*b*) Biological control

(*c*) Physical control (*d*) Mechanical control

51. Which one of the following is a natural factor of pest population reduction?

(*a*) Crop rotation (*b*) Trap crop

(*c*) Clean culture (*d*) Disease of insects

52. Amendment to DIP Act was done in the year

(*a*) 1992 (*b*) 1994

(*c*) 2000 (*d*) 2002

53. The efficacy of pheromonal action depends on

(*a*) Volatility of compound (*b*) Stability in air

(*c*) Rate of diffusion (*d*) All of these

54. National Plant Protection Organization (NPPO) for India is

(*a*) DPPQ and S (*b*) NBPGR

(*c*) NCIPM (*d*) None of these

55. Sticky trap is used to trap the

(*a*) Sucking pests (*b*) Borer pests

(*c*) Shoot feeders (*d*) Fruit borer

56. Factor influencing ETL

(*a*) Market value of crop
(*b*) Degree of injury per insect
(*c*) Crop susceptibility to injury
(*d*) All of these

57. Which is covered by DPPQ and S?

(*a*) CIB and RC
(*b*) Central Insecticide Laboratory
(*c*) Plant Quarantine
(*d*) All of these

58. Supply of germplasm material outside the country or to any Non-Indian is to be done as per the provisions of

(*a*) Biological Diversity Act
(*b*) Varity Act
(*c*) Export and Import Act
(*d*) None of these

59. Compulsory fumigation of imported cotton bales to prevent introduction is to prevent the introduction of

(*a*) Mealy bugs
(*b*) Cotton boll worm
(*c*) Cotton boll weevil
(*d*) Whitefly

60. In mustard early sowing is done to protect the crop from which of the following pest

(*a*) Saw fly
(*b*) Aphids
(*c*) Both (*a*) & (*b*)
(*d*) None of these

61. Which of the following is a cultural method for controlling insects?

(*a*) Crop rotation
(*b*) Quarantine
(*c*) Window screens
(*d*) Sex pheromone traps

62. Only natural control, no insecticides were used during

(*a*) Subsistence phase
(*b*) Exploitation phase
(*c*) Crisis phase
(*d*) Disaster phase

63. Which one is a community level pest management practice?

(*a*) Synchronized sowing
(*b*) Chemical pesticide
(*c*) Mechanical method
(*d*) Cultural method

64. Which of the following insect pest is of domestic quarantine importance?

(*a*) Cotton Cushion Scale
(*b*) Coffee berry borer
(*c*) San Jose Scale
(*d*) All of these

65. Trap crop for IPM of DBM is

(*a*) Tomato
(*b*) Carrot
(*c*) Mustard
(*d*) Cotton

66. Clipping of top portion of rice seedlings reduces the carryover of infestation of

(*a*) Stem borer
(*b*) WBPH
(*c*) GLH
(*d*) Rice bug

67. Which of the following insect pest is not present in India?

(*a*) Cotton boll weevil
(*b*) Tobacco bud worm
(*c*) Hessian fly
(*d*) All of these

68. Transgenic variety of crop generally use to control which insect order

(*a*) Hemiptera
(*b*) Diptera
(*c*) Lepidoptera
(*d*) Coleoptera

69. Early sowing of mustard crop to escapes aphid damage, it is a

(*a*) Cultural control
(*b*) Mechanical control
(*c*) Physical control
(*d*) Biological control

70. Domestic quarantine measures have been implicated against

(*a*) San Jose Scale
(*b*) Fruit fly
(*c*) Mango hoppers
(*d*) Sugarcane scale

71. Which of the following is correct with respect of quantitative losses?

(*a*) Losses of weight
(*b*) Losses of grain due to depredation
(*c*) Losses due to grain mishandling
(*d*) All of these

72. Which of the following is correct with respect of qualitative losses?

(*a*) Presence of insect excreta
(*b*) Discoloration of the grain
(*c*) Bad smell
(*d*) All of these

73. Which of the following comes under direct losses?

(*a*) Injury by chewing insects
(*b*) Injury by internal feeders
(*c*) Both (*a*) & (*b*)
(*d*) None of these

74. Which of the following comes under indirect losses?

(*a*) Making the harvest more difficult
(*b*) Disseminate plant disease
(*c*) Causing contamination
(*d*) All of these

75. Method of mass trapping of insects at night comes under

(*a*) Mechanical control
(*b*) Physical control
(*c*) Cultural control
(*d*) Biological control

76. Recent approaches of insect control is

(*a*) Biological control (*b*) Chemical control

(*c*) Mechanical control (*d*) Integrated Pest Management

77. Making trench for the pest management comes under

(*a*) Legal control (*b*) Cultural control

(*c*) Physical control (*d*) Mechanical control

78. Host plant resistance comes under

(*a*) Legal control (*b*) Cultural control

(*c*) Genetical control (*d*) Mechanical control

79. IPM be supposed to applied on

(*a*) Resistant pest (*b*) Susceptible pest

(*c*) Economic Threshold Level (*d*) Any loss of crops by pests

80. Flooding of the field is effective control of

(*a*) Aphid (*b*) Thrips

(*c*) Termites (*d*) Budworms

81. Quarantine measures have been implicated against

(*a*) Root knot nematode (*b*) Golden cyst nematode

(*c*) Stunt nematode (*d*) Reniform nematode

82. Natural insect control factors are

(*a*) Physical barriers (*b*) Climatic factors

(*c*) Natural enemies and insect disease (*d*) All of these

83. Economic threshold level is always

(*a*) Equal to EIL (*b*) More than EIL

(*c*) Lower than EIL (*d*) None of these

84. The aim of IPM is

(*a*) Keeping pest population below economic threshold level

(*b*) Keeping pest population below economic injury level

(*c*) Keeping pest population below general equilibrium level

(*d*) None of these

85. Name of the pest which has migrated from India to other countries

(*a*) *Sitotroga cerealella* (*b*) *Sitophilus oryzae*

(*c*) *Trogoderma granarium* (*d*) None of these

86. Light trap is used for which type of insect pests

(a) Diurnal (b) Nocturnal

(c) Both (a) & (b) (d) None of these

87. A better pest management depends upon

(a) Use of resistant variety (b) Biological control

(c) Uses of pesticide (d) Interlacing all known control methods

88. Latest method of insect pest control is

(a) Eradication (b) Control

(c) Pest management (d) Both (a) & (b)

89. Flooding of the fields will help in control of

(a) White grubs (b) Termites

(c) Nematodes (d) All of these

90. Stripping of sugarcane leaves minimizes the attack of

(a) Early shoot borer (b) Mealy bugs

(c) Scale insect (d) Top shoot borer

91. Alternate wetting and drying of paddy fields is done for the management of

(a) Yellow stem borer (b) Gundhi bug

(c) Brown Plant Hopper (d) Leaf folder

92. Alley cropping is cultural practice for the management of

(a) Yellow stem borer (b) Gundhi bug

(c) Brown Plant Hopper (d) Leaf folder

93. Passing of rope over the crop helps to manage

(a) Yellow stem borer (b) Rice case worm

(c) Brown Plant Hopper (d) Leaf folder

94. In the reduction of a pest population, the balance of the natural enemies is maintained in

(a) Pest management (b) Pest control

(c) Pest eradication (d) None of these

95. Killing of which stage of insect is the most effective control measures for early stage gregarious pests

(a) Egg (b) Larva

(c) Pupa (d) Adult

96. Indiscriminate use of pesticides and cultivation of high yielding crop varieties often leads to

(a) A minor disease or insect pest problem becoming serious threat

(b) Destruction of natural enemies of pests

(c) Appearance of pesticides resistant pest strains

(d) All of these

97. Light trapping of insects is comes under the

(a) Preventive method (b) Cultural method

(c) Mechanical method (d) Legal method

98. Light traps are used for the collection of__________insects.

(a) Geotropic (b) Chemotropic

(c) Negatively phototropic (d) Positively phototropic

99. From the quarantine point of view which is the land frontier

(a) Kalimpong (b) Kolkata

(c) Kandla (d) New Delhi

100. Without chemical treatment termite control can also be done by

(a) Tillage (b) Clean cultivation

(c) Irrigation (d) Trap cropping

Fill in the Blanks

1. Leaving a population of pest for feed of their natural enemy is known as__________.

2. The ETL always represents a pest density lower than that of__________.

3. __________implies that an insect with resistance to one insecticide is able to resist other insecticides.

4. A situation where a population of insect, after having suppressed, rebounds to numbers greater than before suppression occurred is known as__________.

5. Different populations of an insect species that vary in their virulence to a cultivar are referred as__________.

Match the Following

1. **Match the method of management and example**

(a) Physical control	i) Hand picking of insects
(b) Mechanical control	ii) Use of predator and parasitoids
(c) Biological control	iii) Alternate drying and wetting of fields
(d) Cultural control	iv) Deep summer ploughing

2. **Match the pests with their trap crops**

(a) Leaf miner	i) Castor
(b) Gram pod borer	ii) Soybean
(c) Diamond backed moth	iii) Mustard
(d) Tobacco caterpillar	iv) African marigold

3. **Match the following control measures with respective pests**

(a) Alternate drying and wetting of paddy fields	i) Brown Plant Hopper
(b) Coating grains with vegetable oils	ii) Aphids and Whitefly
(c) Use of yellow sticky traps	iii) Stored product insects
(d) Passing of rope over paddy fields	iv) Paddy Case Worm

4. **Match the following management practices and respective insects**

(a) Banding of mango trees	i) Rat
(b) Use of tin band in coconut	ii) White grubs and termites
(c) Use of iron hook	iii) Mealy bug
(d) Flooding of fields	iv) Rhinoceros beetle

Answers

Multiple Choice Questions

1	(d)	21	(a)	41	(a)	61	(a)
2	(d)	22	(d)	42	(a)	62	(a)
3	(a)	23	(d)	43	(a)	63	(a)
4	(c)	24	(c)	44	(a)	64	(d)
5	(a)	25	(d)	45	(c)	65	(c)
6	(a)	26	(a)	46	(c)	66	(a)
7	(b)	27	(d)	47	(b)	67	(d)
8	(a)	28	(a)	48	(c)	68	(c)
9	(a)	29	(a)	49	(c)	69	(a)
10	(a)	30	(a)	50	(b)	70	(a)
11	(a)	31	(a)	51	(d)	71	(d)
12	(a)	32	(a)	52	(a)	72	(d)
13	(b)	33	(d)	53	(d)	73	(c)
14	(c)	34	(d)	54	(a)	74	(d)
15	(d)	35	(a)	55	(a)	75	(a)
16	(b)	36	(a)	56	(d)	76	(d)
17	(a)	37	(a)	57	(d)	77	(d)
18	(a)	38	(c)	58	(a)	78	(b)
19	(a)	39	(a)	59	(c)	79	(d)
20	(d)	40	(a)	60	(b)	80	(c)

81	(b)	86	(b)	91	(c)	96	(d)
82	(d)	87	(d)	92	(c)	97	(c)
83	(b)	88	(c)	93	(b)	98	(d)
84	(a)	89	(d)	94	(a)	99	(a)
85	(c)	90	(b)	95	(a)	100	(c)

Fill in the Blanks

1. Pest residue
2. EIL
3. Cross resistance
4. Resurgence
5. Biotypes

Match the Following

1. a) - iii, b) - i, c) - ii, d) - iv
2. a) - ii, b) - iv, c) - iii, d) - i
3. a) - i, b) - iii, c) - ii, d) - iv
4. a) - iii, b) - i, c) - iv, d) – ii

Chapter 10

Pesticide Application Equipments

Multiple Choice Questions

1. **The safest method of disposing off the left over pesticide is disposal by**

   (a) Pouring in the drain (b) Pouring in a pond

   (c) Pouring in a moving stream (d) Burying in the soil

2. **The liquid formulation of insecticides are applied with the help of a**

   (a) Planter (b) Duster

   (c) Drill (d) Sprayer

3. **Knapsack, Bucket pump, Stirrup Pumps are__________type of sprayer.**

   (a) Low volume (b) High volume

   (c) ULV (d) Dusting

4. **In Foot sprayers, Rocking sprayers, Knapsack sprayers, Bucket pump, Stirrup Pumps__________is used.**

   (a) Hydraulic energy (b) Thermal energy

   (c) Electrical energy (d) Wind energy

5. **Which of the following protective measures are necessary for person manually dusting pesticide on crop in uncertain wind condition?**

   (a) Wearing plastic aprons (b) Use of goggles

   (c) Use of nose filter (d) All of these

6. Mist blowers, controlled droplet applicators, electrically operated blower-cum sprayers are

(*a*) Low volume sprayer (*b*) High volume sprayer
(*c*) ULV sprayer (*d*) Duster

7. ULV stands for

(*a*) Ultra Liter Volume (*b*) Ultra Low Volume
(*c*) Ultra Light Volume (*d*) Ultra Lit Volume

8. In ULV sprayer__________energy is utilized.

(*a*) Gaseous (*b*) Centrifugal
(*c*) Electrical (*d*) All of these

9. 1 μ equal to

(*a*) 1 mm (*b*) 0.1 mm
(*c*) 0.01 mm (*d*) 0.001 mm

10. In swig fog machine, smoke generators and vapour generators__________ energy is utilized.

(*a*) Gaseous (*b*) Centrifugal
(*c*) Electrical (*d*) Thermal

11. Vmd stands for

(*a*) Volume Mean Diameter (*b*) Volume Median Diameter
(*c*) Value Mean Dose (*d*) Volume Median Dose

12. Which one of the following remedies is to be suggested in case of failure of a manually operated sprayer to retain pressure?

(*a*) Tightening of loose nuts and clamps and replacement of gasket
(*b*) Tightening of the lid of the tank, replacement of gasket, if necessary
(*c*) Opening the nozzle and cleaning its various parts especially the orifice
(*d*) Straightening the plunger rod if found bent

13. Vmd of vapour is ranged between__________μm.

(*a*) 0.001 - 0.01 (*b*) 0.01 - 1
(*c*) 1 - 50 (*d*) 51 - 100

14. Vmd of smoke is ranged between__________μm.

(*a*) 0.001 - 0.01 (*b*) 0.01 - 1
(*c*) 1 - 50 (*d*) 51 - 100

15. The quantity of liquid insecticide with 25 per cent active ingredient required for preparing 500 liters of the spray fluid of 0.25 per cent strength is

(*a*) 1.25 liter (*b*) 2.25 liter

(*c*) 5.00 liter (*d*) 10.00 liter

16. Vmd of aerosol/fog is ranged between__________µm.

(*a*) 0.001 - 0.01 (*b*) 0.01 - 1

(*c*) 1 - 50 (*d*) 51 - 100

17. Droplet size or Vmd of mist spray is ranged between__________µm.

(*a*) 0.001 - 0.01 (*b*) 0.01 - 1

(*c*) 1 - 50 (*d*) 51 - 100

18. Droplet size or Vmd of fine spray is ranged between__________µm.

(*a*) 51 - 100 (*b*) 101 - 200

(*c*) 201 - 400 (*d*) > 400

19. Droplet size or Vmd of medium spray is ranged between__________µm.

(*a*) 51 - 100 (*b*) 101 - 200

(*c*) 201 - 400 (*d*) > 400

20. Droplet size or Vmd of coarse spray is ranged between__________µm.

(*a*) 51 - 100 (*b*) 101 - 200

(*c*) 201 - 400 (*d*) > 400

21. To control flying insects__________micron spray droplet are used.

(*a*) 5 - 10 (*b*) 10 - 50

(*c*) 30 - 150 (*d*) 100 - 300

22. To control plant diseases and insect on surfaces__________micron spray droplet are used.

(*a*) 10 - 15 (*b*) 20 - 25

(*c*) 30 - 150 (*d*) 100 - 300

23. Which one among the following is the pneumatic sprayer?

(*a*) Foot sprayer (*b*) Stirrup sprayer

(*c*) Hand compression sprayer (*d*) Rocker sprayer

24. To control weeds__________micron spray droplet are used.

(*a*) 10 - 15 (*b*) 20 - 25

(*c*) 30 - 150 (*d*) 100 - 300

25. In case of rotary type of duster the air current is developed by

(a) Bellows
(b) Fans
(c) Self propelling jacket
(d) Self propelling blades

26. In case of Controlled Droplet Applicator (CDA) the spinning disc rotates at a very high speed i.e.

(a) < 2000 rpm
(b) 2000 - 4000 rpm
(c) 6000 - 9000 rpm
(d) > 10000 rpm

27. Nozzle is a critical part of any

(a) Duster
(b) Sprayer
(c) Dibbler
(d) Driller

28. Nozzle works as

(a) Regulate flow
(b) Atomize the mixture into droplets
(c) Disperse the spray in a desirable pattern
(d) All of these

29. The optimum pressure to produce spray droplet is

(a) 800 psi
(b) 128 psi
(c) 400 psi
(d) 218 psi

30. The optimum pressure to produce spray droplet is

(a) 800 pscm
(b) 128 pscm
(c) 400 pscm
(d) 218 pscm

31. Nozzles determine the rate of pesticide distribution at a particular

(a) Temperature
(b) Humidity
(c) Pressure
(d) Light

32. Cap of nozzle holds with body

(a) Strainer
(b) Orifice plate
(c) Washer and Swirl phase
(d) All of these

33. Holes or openings on the strainer are very small which prevents the entry of

(a) Bigger size particles
(b) Smaller size particles
(c) Spray mixture
(d) Water

34. The use of flood jet nozzle is required for

(a) Minimizing the drift of chemicals
(b) Increasing the droplet size
(c) Quick delivery of spray fluid
(d) Better adhesion of spray fluid

35. Pneumatic sprayers have been used in pest control for many years. A pneumatic sprayer is a

(*a*) Operates using wind or air (*b*) Contains poisonous pesticide
(*c*) A knapsack sprayer (*d*) Operates automatically

36. Which part of the nozzle of a sprayer prevents the leakage of spray fluid?

(*a*) Cap (*b*) Washer
(*c*) Swirl phase (*d*) Orifice plate

37. Definite characteristic spray patterns are given by

(*a*) Cap (*b*) Washer
(*c*) Swirl phase (*d*) Orifice plate

38. Nozzle flow rate is a function of

(*a*) Cap (*b*) Washer
(*c*) Swirl phase (*d*) Orifice plate

39. Droplet size of the spray liquid determined by

(*a*) Design of the nozzle (*b*) Pressure
(*c*) Sprayer body (*d*) Both (*a*) & (*b*)

40. Nozzle strainers are equipped with

(*a*) Check valves (*b*) Washer
(*c*) Swirl phase (*d*) Orifice plate

41. Check valves cause a pressure drop of

(*a*) 5 to 10 psi (*b*) 11 to 15 psi
(*c*) 16 to 20 psi (*d*) 21 - 25 psi

42. Spray pattern has two basic characteristics

(*a*) The spray angle and shape of the pattern
(*b*) The spray angle and quality of water
(*c*) Shape of the pattern and quality of water
(*d*) None of these

43. Generally agricultural nozzles have an angle between

(*a*) 20 - 45 degrees (*b*) 20 - 60 degrees
(*c*) 65 - 120 degrees (*d*) 120 - 180 degrees

44. Considerable point for good spray coverage

(*a*) Nozzle discharge angle
(*b*) Nozzle distance from spray surface
(*c*) Nozzle spacing on the boom
(*d*) All of these

45. Which of the following is a function of a nozzle?

(a) Regulate flow (b) Atomize mixture into droplets

(c) Disperse the spray in desirable pattern (d) All of these

46. The common principle involved in the operation of duster is to utilize

(a) Droplets (b) Smokes

(c) The air blast (d) Temperature

47. Granules are normally applied in the

(a) Seed furrows (b) Leaf whorls

(c) Old branches (d) Both (a) & (b)

48. To control the pest of fruit trees (upto 4 metres height)__________are used.

(a) Knapsac sprayers (b) Mist blowers

(c) Foot sprayers (d) Rocker sprayers

49. To spraying coconut and palm trees (nearly 5 m height)__________are used.

(a) Knapsac sprayers (b) Mist blowers

(c) Foot sprayers (d) Rocker sprayers

50. Pressure maintain in knapsack sprayer is

(a) 3 - 5 kg/sq cm (b) 17 - 21 kg/sq cm

(c) 36 kg/sq cm (d) 40 kg/sq cm

51. A fine aerosol, droplet size ranging between 1 - 50 μ is known as

(a) Mist (b) Fog

(c) Vapour (d) Smoke

52. Where spray liquid volume required in between 5 and 400 litre/ha__________spray techniques are employed?

(a) Low volume (b) High volume

(c) ULV (d) Dusting

53. Less than 5 litre/ha spray volume is applied using__________technique.

(a) Low volume (b) High volume

(c) ULV (d) Dusting

54. Which one of the following can be classified as hydraulic energy sprayer?

(a) Mist blower

(b) Controlled drip applicator sprayer

(c) Knapsack sprayer

(d) Wheel barrow sprayer

55. Propelling of dusts into the blower of a duster is made by

(*a*) Nozzle (*b*) Valves

(*c*) Agitators (*d*) Hopper

56. Which one of the following is used for fogs composed of very fine droplets?

(*a*) Gaseous energy nozzle (*b*) Centrifugal energy nozzle

(*c*) Kinetic energy nozzle (*d*) Thermal energy nozzle

57. The quantity of Chlorpyriphos 20 per cent EC required for treating an area which needs 1000 liters of spray fluid at 0.05 per cent (a.i.) strength is

(*a*) 20 x 0.05/1000 liters (*b*) 0.5 x 1000/1000 liters

(*c*) 1000 x 0.05/20 liters (*d*) 20 x 1000/0.05 liters

58. Which of the following statement is true?

(*a*) Flat fan spray nozzles are widely used for broadcast spraying of herbicides and some insecticides

(*b*) Flooding fan nozzles are used for applying herbicides mixtures and liquid fertilizers

(*c*) Full cone nozzles is a low drift nozzle and is used to apply soil incorporated herbicides

(*d*) All of these

59. Paddle pump sprayer can spray up to the height of

(*a*) 10 feet (*b*) 15 feet

(*c*) 20 feet (*d*) 30 feet

60. The duster hopper should be filled with dust to its capacity

(*a*) Full (*b*) 1/3

(*c*) 1/2 (*d*) 3/4

61. The size (in micrometers in diameter) of droplets produced by high volume sprayers should most appropriately be

(*a*) 200 to 500 (*b*) 100 to 150

(*c*) 81 to 100 (*d*) 30 to 40

62. Nozzle is a part of

(*a*) Cyanogas pump (*b*) Rotary duster

(*c*) Hand compressor sprayer (*d*) None of these

63. Name of the Nozzle which is used for producing mist

(*a*) Blast nozzle (*b*) Rotary nozzle

(*c*) Kinetic energy nozzle (*d*) Annular nozzle

64. The nozzle used for spraying herbicides is

(a) Shear nozzle (b) Flat fan
(c) Hollow cone (d) None of these

65. Example of high volume sprayer

(a) Knapsack sprayer (b) Foot sprayer
(c) Rocker sprayer (d) All of these

66. Part of sprayer breaks the spray liquid into fine droplets

(a) Pump (b) Agitator
(c) Nozzle (d) None of these

67. In case of field crops if spraying per hectare is less than 50 litres of water it is termed as

(a) High volume spray (b) Low volume spray
(c) Ultra low volume spray (d) Very low volume spray

68. If spraying contains per hectare of field crops is 50 to 200 litres of water it is termed as

(a) High volume spray (b) Low volume spray
(c) Ultra low volume spray (d) None of these

69. If spraying per hectare field crops is 200 - 600 litres of water is termed as

(a) High volume spray (b) Low volume spray
(c) Ultra low volume spray (d) Medium volume spray

70. Which one of the following types of spraying is best done by manually operated knapsack having flat-fan nozzle?

(a) High volume (b) Low volume
(c) Ultra low volume (d) Semi - low volume

71. Which type of pump is used in a hand compression sprayer?

(a) Rotary pump (b) Air or Pneumatic
(c) Piston pump (d) All of these

72. Which one is an example of high volume sprayer?

(a) Hand sprayer (b) ULV sprayer
(c) Foot sprayer (d) Stirrup pumps

73. Which of the following nozzles are generally considered most suitable for undertaking spraying of insecticide spray by manually operated sprayers?

(a) Solid stream (b) Flat fan
(c) Hollow cone (d) Solid cone

74. Most of soil fumigants are applied with

(a) Duster (b) Sprayer

(c) Soil injector (d) Dispenser

75. Triple action nozzle provides one of the following spray patterns

(a) Hollow cone + Solid cone + Jet

(b) Hollow cone + Jet + Flat fan

(c) Hollow cone + Jet

(d) Hollow cone + Solid cone + Flat fan

76. The device fitted at the end of the spray lance and is used for automization is

(a) Boom (b) Hose

(c) Cut-off-device (d) Nozzle

77. Which one of the following sprayers will cause more loss of spray droplets by drift?

(a) Low volume sprayer (b) High volume sprayer

(c) Ultra low volume sprayer (d) Duster

78. In case of rotary type of duster the air current is developed by

(a) Bellows (b) Fans

(c) Self propelling jacket (d) Self propelling blades

79. If one gram of a pesticide formulation containing 50 per cent active ingredient is mixed with 1 litre of water, what will be the concentration of the active ingredient in the spray fluid

(a) 0.5 ppm (b) 5.0 ppm

(c) 50 ppm (d) 500 ppm

80. In solid cone nozzle the solid cone of liquid is formed due to

(a) Bigger size of the cap (b) Bigger size of the orifice plate

(c) Central hole of swirl plate (d) Small size of the strainer

81. In which one of the following situations is the aerial spraying insecticide effective and advisable?

(a) Presence of different kinds of pest population in the crop

(b) Uniform cultivation of the same crop available in the area

(c) Heavy attack of sucking type of pests

(d) Presence of different types of crops in the area

82. If spraying per hectare field crops is more than 600 litres of water is used it termed as

(a) High volume spray (b) Low volume spray

(c) Ultra low volume spray (d) Medium volume spray

83. The nozzle in which the spray fluid is fed into the air stream within the nozzle and spitted into spray droplets by the velocity of air is called

(*a*) Disc nozzle (*b*) Annular nozzle

(*c*) Shear nozzle (*d*) Low volume nozzle

84. The energy is utilized for making fog, smoke and vapour sprays

(*a*) Thermal (*b*) Gaseous

(*c*) Centrifugal (*d*) Hydraulic

85. The equipment which atomizes the spray fluid into small droplets and ejects it some force for proper distribution is called

(*a*) Duster (*b*) Sprayer

(*c*) Dibbler (*d*) Boom

86. In a sprayer, the metal rod on which nozzle is attached is called

(*a*) Nozzle (*b*) Boom

(*c*) Spray lance (*d*) Bellows

87. Spraying is usually done in the__________hours.

(*a*) Noon hours (*b*) Morning hours

(*c*) Evening hours (*d*) Night hours

88. Dusting is usually done in the__________hours.

(*a*) Noon hours (*b*) Morning hours

(*c*) Evening hours (*d*) Night hours

89. Pesticides are better applied to field crops

(*a*) In the morning only (*b*) In the evening only

(*c*) During mid noon (*d*) Either a or b

90. ULV application is preferred when__________areas to be treated.

(*a*) Small (*b*) Large

(*c*) Medium (*d*) All of these

Fill in the Blanks

1. __________are used for application of powders or dusts.

2. __________are used for application of solutions.

3. For applications of gases__________are used.

4. In__________the air or other gas is used for producing spray droplets.

5. For application of herbicides usually__________nozzle is used.

6. Flat fan nozzle is also called__________.

7. In gaseous-energy nozzle the air or other gas is used for producing spray droplets, such type of nozzle is also called__________.

Match the Following

1. Match the spray with their droplet size

(a) Vapour	i) 0.001 - 0.01 µm
(b) Smoke	ii) 1 - 50 µm
(c) Fog/Aerosol	iii) 51 - 100 µm
(d) Mist	iv) 0.01 - 1 µm

2. Match the spray with their droplet size

(a) Mist	i) > 400 µm
(b) Fine spray	ii) 51 - 100 µm
(c) Medium spray	iii) 201 - 400 µm
(d) Coarse spray	iv) 101 - 200 µm

3. Match the parts of nozzle and their works

(a) Cap	i) Definite characteristics spray pattern
(b) Washer	ii) Prevents leakage
(c) Swirl phase	iii) Holds the body
(d) Orifice plate	iv) Maintain nozzle flow rate

4. Match the sprayer with area of use

(a) Knapsack sprayer	i) Plantation crops
(b) Pneumatic hand sprayer	ii) Kitchen garden
(c) Foot pump sprayer	iii) Field crops
(d) Rocker sprayer	iv) Fruit trees

5. Match the equipments with their type

(a) Knapsack sprayer	i) ULV sprayer
(b) Rocker sprayer	ii) Fumigator
(c) Battery operated control drop applicator	iii) High volume sprayer
(d) Soil injector	iv) Low volume sprayer

Answers

Multiple Choice Questions

1	(d)	21	(b)	41	(a)	61	(a)
2	(d)	22	(c)	42	(a)	62	(c)
3	(b)	23	(c)	43	(c)	63	(b)
4	(a)	24	(d)	44	(d)	64	(b)
5	(d)	25	(b)	45	(d)	65	(a)
6	(c)	26	(c)	46	(c)	66	(c)
7	(b)	27	(b)	47	(d)	67	(d)
8	(d)	28	(d)	48	(c)	68	(b)
9	(d)	29	(a)	49	(d)	69	(d)
10	(d)	30	(b)	50	(a)	70	(a)
11	(b)	31	(c)	51	(b)	71	(c)
12	(b)	32	(d)	52	(a)	72	(d)
13	(a)	33	(a)	53	(c)	73	(c)
14	(b)	34	(a)	54	(c)	74	(c)
15	(c)	35	(c)	55	(c)	75	(a)
16	(c)	36	(b)	56	(d)	76	(d)
17	(d)	37	(c)	57	(c)	77	(c)
18	(b)	38	(d)	58	(d)	78	(b)
19	(c)	39	(d)	59	(c)	79	(d)
20	(d)	40	(a)	60	(d)	80	(c)

81	(c)	84	(a)	87	(c)	90	(b)
82	(a)	85	(b)	88	(b)		
83	(c)	86	(c)	89	(d)		

Fill in the Blanks

1. **Dusters**
2. **Sprayers**
3. **Fumigators**
4. **Gaseous energy nozzle**
5. **Kinetic-energy**
6. **Fish tail nozzle**
7. **Twin-fluid nozzle**

Match the Following

1. **a) - i, b) - iv, c) - ii, d) - iii**
2. **a) - ii, b) - iv, c) - iii, d) - i**
3. **a) - iii, b) - ii, c) - i, d) - iv**
4. **a) - iii, b) - ii, c) - iv, d) - i**
5. **a) - iii, b) - iv, c) - i, d) - ii**

Chapter 11
Pests of Crops and their Management

Multiple Choice Questions

1. **Which pest of paddy belongs to the family Coreidaae?**

   (a) Rice bug (b) WBPH
   (c) BPH (d) None of these

2. **Maximum losses in paddy field due to attack with same population size stem borer is in which stage**

   (a) Panicle initiation (b) Active tillering
   (c) Flowering (d) Seedling

3. **Maximum loss to paddy is caused by YSB in which of the following state**

   (a) Rajasthan (b) Andhra Pradesh
   (c) Goa (d) Jammu and Kashmir

4. **Which of the following is a dipteran pest of rice?**

   (a) Yellow Stem Borer (b) Brown Plant Hopper
   (c) Rice Hispa (d) Rice Whorl maggot

5. **Which of the following is a monophagous pest of rice?**

   (a) Yellow stem borer (b) Gundhi bug
   (c) Case worm (d) Leaf folder

6. **Dead heart of rice is caused by**

   (a) Yellow stem borer (b) Gundhi bug
   (c) Case worm (d) Leaf folder

7. **Onion leaf symptoms of rice is caused by**
   (*a*) Yellow stem borer (*b*) Gundhi bug
   (*c*) Case worm (*d*) Gall midge

8. **White parallel streaks along the leaf axis of rice is caused by**
   (*a*) Yellow stem borer (*b*) Rice hispa
   (*c*) Case worm (*d*) Leaf folder

9. **Cut leaf bit floating in water is characteristic symptom of**
   (*a*) Yellow stem borer (*b*) Gundhi bug
   (*c*) Case worm (*d*) Leaf folder

10. **Marginal rolling in rice seedling is damage symptom**
    (*a*) Yellow stem borer (*b*) Gundhi bug
    (*c*) Case worm (*d*) Rice thrips

11. **Individual rice grain become white and chaffy due to the attack of**
    (*a*) Yellow stem borer (*b*) Gundhi bug
    (*c*) Case worm (*d*) Leaf folder

12. **Increased silica content affects the mandibular development of**
    (*a*) Yellow stem borer (*b*) Gundhi bug
    (*c*) Case worm (*d*) Leaf folder

13. **In rice, clipping of seedling tip before transplanting is done to remove egg masses of**
    (*a*) Yellow stem borer (*b*) Gundhi bug
    (*c*) Brown Plant Hopper (*d*) Leaf folder

14. **Mirid bug, *Cyrtorhinus lividipennis* is a natural enemy of**
    (*a*) Yellow stem borer (*b*) Gundhi bug
    (*c*) Brown Plant Hopper (*d*) Rice Gall Midge

15. **In paddy, the elongation of leaf sheath, when infested by gall midge is due to a chemical known as**
    (*a*) Cecimogen (*b*) Ceciden
    (*c*) Orseolin (*d*) Cecidogen

16. **White ear head in paddy is caused by**
    (*a*) Yellow stem borer (*b*) Green leaf hopper
    (*c*) Gall fly (*d*) None of these

17. **Yellow stem borer hibernates as**
    (*a*) Egg (*b*) Larva
    (*c*) Pupa (*d*) Adult

18. ***Heiroglyphus nigrorepletus*** **is the scientific name of**

(a) Rice grasshopper (b) Gundhi bug

(c) Pink bollworm (d) Pink stem borer

19. "Hopper Burn" appears in

(a) Rice (b) Wheat

(c) Sugarcane (d) Maize

20. Sorghum aphid is scientifically known as

(a) *Hysteraneura setariae* (b) *Melanaphis sacchari*

(c) *Brevenia* spp. (d) Both (a) & (b)

21. Older rice leaves fasten along with the longitudinal margins is due to

(a) *Chilo incertulus* (b) *Cnaphalocrocis medinalis*

(c) *Dicladispa armigera* (d) *Contarinia sorghicola*

22. Borer of ragi is scientifically known as

(a) *Chilo suppresalis* (b) *Sesamia inferens*

(c) *Chilo partellus* (d) *Scirpophaga nivella*

23. Sorghum shoot fly lay eggs on

(a) Flowers (b) Upper surface of leaves

(c) Lower surface of leaves (d) Stem

24. Ear head midge is scientifically known as

(a) *Calocoris anguistata* (b) *Contarinia sorghicola*

(c) *Orseolia oryzae* (d) None of these

25. Gujhia weevil is an important pest of

(a) Paddy (b) Maize

(c) Wheat (d) Gram

26. __________is a serious pest of Jowar and Bajra.

(a) Stem borer (b) Midge

(c) White borer (d) Aphid

27. Maize stem borer hibernates in

(a) Egg stage (b) Larval stage

(c) Pupal stage (d) Adult stage

28. In sorghum, dead hearts with more side tiller is characteristic symptom of

(a) Shoot fly (b) Stem borer

(c) Sorghum midge (d) Ear head bug

29. Sorghum grain formation has been stopped due to the feeding of

(a) *Contarinia sorghicola* (b) *Atherigona soccata*
(c) *Chilo partellus* (d) *Calocoris angustata*

30. Which one of the following pests attacks rice crop during milking stage?

(a) *Chilo partellus* (b) *Plutella xylostella*
(c) *Pericallia ricini* (d) *Mythimna separeta*

31. The shoot fly females lay white, elongated, cigar shaped eggs singly on

(a) The undersurface of the leaves, parallel to the midrib
(b) Stem
(c) Upper surface of the leaves
(d) Flowers

32. Attack of which insects may be recognized by unpleasant smell from rice field

(a) Grasshopper (b) Stem borer
(c) Gundhi bug (d) None of these

33. Notorious pest of jowar and bajra is

(a) Sorghum midge (b) Stem borer
(c) Wheat borer (d) Aphid

34. Which of the following is used for the management of shoot fly in sorghum?

(a) Early sowing
(b) High seed rate and fish meal trap
(c) Application of granular insecticides in whorls
(d) All of these

35. Which insect feeds on silk and developing cobs of maize?

(a) *Helicoverpa armigera* (b) *Mythimna separeta*
(c) *Chilo zonellus* (d) None of these

36. Which of the following pests attacks wheat crop?

(a) Root aphid (b) Termites
(c) Stem borer (d) All of these

37. Most notorious pest of wheat is

(a) Cut worm (b) Termite
(c) Thrips (d) Aphid

38. Dead heart in the central shoot of wheat is caused by

(*a*) Stem borer (*b*) Root borer

(*c*) Root aphid (*d*) None of these

39. Presence of large number of ants at the base of the finger millet plant is characteristic symptom of

(*a*) Ragi root aphid (*b*) Stem borer

(*c*) Termite (*d*) None of these

40. Dead heart and chaffy ear head in ragi is caused due to

(*a*) Ragi root aphid (*b*) Stem borer

(*c*) Termite (*d*) None of these

41. Stem borer are common to which of the following groups

(*a*) Rice, Wheat, Sugarcane (*b*) Mustard, Sesame, Safflower

(*c*) Gram, Mung, Arhar (*d*) Bhindi, Tomato, Brinjal

42. In mustard early sowing is recommended to control

(*a*) Saw fly (*b*) Aphid

(*c*) Flea beetle (*d*) Painted bug

43. In sorghum which of the following pest attacks the crop exclusively after one month is

(*a*) Shoot fly (*b*) Stem borer

(*c*) Sorghum midge (*d*) Ear head bug

44. The basis of antibiosis in corn against *Helicoverpa zea* is

(*a*) Lethal silk factor (*b*) Nutritional imbalance

(*c*) Presence of growth inhibitor (*d*) Deficiency of nutrients

45. The basis of antibiosis in maize European corn borer is

(*a*) Nutritional imbalance

(*b*) Presence of growth inhibitors MBOA and DIMBOA

(*c*) Deficiency of nutrient

(*d*) None of these

46. Which pest of maize belongs to family Pyralidae?

(*a*) *Chilo partellus* (*b*) *Sesamia inference*

(*c*) *Mythimna separeta* (*d*) None of these

47. *Atherigona socata* is commonly known as

(*a*) Shoot fly (*b*) Cut worm

(*c*) Pod borer (*d*) Army worm

48. Occurrence of white parallel streaks along leaf axis in rice is due to

(*a*) *Scirpophaga incertulus* (*b*) *Chilo partellus*

(*c*) *Cnaphalocrocis medinalis* (*d*) *Dicladespa armigea*

49. Hopper burn symptoms of rice is caused by

(*a*) Brown Plant Hopper (*b*) White Backed Plant Hopper

(*c*) Both (*a*) & (*b*) (*d*) None of these

50. Cutting/nibbling of rice plant is caused by

(*a*) Plant hopper (*b*) Gundhi bug

(*c*) Rice grasshopper (*d*) Cut worms

51. Jowar stem borer belongs to the family

(*a*) Coccidae (*b*) Carambidae

(*c*) Noctuidae (*d*) Arctidae

52. Clipping of young rice seedlings to destroy egg mass of

(*a*) Leaf folder (*b*) Yellow stem borer

(*c*) Grasshopper (*d*) Case worm

53. Dead heart in wheat is characteristic symptoms of

(*a*) Stem borer (*b*) Termite

(*c*) Aphids (*d*) Thrips

54. Pest of maize feed on silk and developing cob

(*a*) *Chilo suppresalis* (*b*) *Helicoverpa armigera*

(*c*) *Chilo partellus* (*d*) *Scirpophaga nivella*

55. *Rophalosyphum maeidis* is commonly known as

(*a*) Sawfly (*b*) Aphid

(*c*) Housefly (*d*) Thrips

56. Which of the following aphid feeds on pulses?

(*a*) *Rhopalosiphum maidis* (*b*) *Aphis craccivora*

(*c*) *Myzus persicae* (*d*) *Lipaphis erysimi*

57. Circular hole of gram pod and plugged by head/body of larvae is major damaging symptoms of

(*a*) Cut worm (*b*) Gram pod borer

(*c*) Blister beetle (*d*) Aphids

58. Pest of pulses attack on both in field and store

(*a*) Cut worm (*b*) Gram pod borer

(*c*) Blister beetle (*d*) Pulse beetle

59. *Aceria cajani* **belongs to family**
   (a) Eriophyidae
   (b) Aphidae
   (c) Bruchidae
   (d) Tettanychus

60. **Withering, patches and drying of groundnut is caused by**
   (a) *Spilosoma oblique*
   (b) *Holotrichia consanguinea*
   (c) *Helicoverpa armigera*
   (d) None of these

61. *Helicoverpa armigera* **damage the gram crop at which stage**
   (a) Pod formation stage
   (b) Maturity stage
   (c) Germinating stage
   (d) Sprouting stage

62. **Plume moth is a pest of**
   (a) Bengal gram
   (b) Green gram
   (c) Black gram
   (d) All of these

63. **The damaging stage of gram cut worm is**
   (a) Nymph
   (b) Caterpillar
   (c) Adult
   (d) All of these

64. *Athalia lugens proxima* **belongs to the order**
   (a) Diptera
   (b) Lepidoptera
   (c) Hemiptera
   (d) Hymenoptera

65. **Groundnut bruchid attacks**
   (a) Kernels
   (b) Pods
   (c) Both (a) & (b)
   (d) None of these

66. **Pin head sized bore hole near the septum and shriveling of developing pulse seeds is caused by**
   (a) Gram pod borer
   (b) Plume moth
   (c) Pod fly
   (d) Pod bug

67. **The noctuid which is common pest to both pigeonpea and chickpea is**
   (a) Pod fly
   (b) *Earias vitella*
   (c) *Helicoverpa armigera*
   (d) Plume moth

68. **Wilting of groundnut plants in patches with damaged roots is caused by**
   (a) White grub
   (b) Leaf miner
   (c) Termites
   (d) All of these

69. White grub beetles lay eggs

(*a*) On the upper side of the leaf

(*b*) Singly in loose soil up to the depth of 10 cm

(*c*) Near the root at the level of ground

(*d*) Within the floral parts

70. Webbing of leaves in groundnut is a characteristic symptom of

(*a*) Leaf miner (*b*) Thrips

(*c*) Leafhoppers (*d*) None of these

71. In groundnut, typical 'V' shaped yellowing at the tip of the leaflets is caused by

(*a*) Leaf miner (*b*) Thrips

(*c*) Leafhoppers (*d*) None of these

72. Groundnut leafminer pupates in

(*a*) Soil (*b*) Between the webbed leaflets

(*c*) Flower (*d*) None of these

73. Scarification of groundnut pods is caused by

(*a*) Earwig (*b*) Termites

(*c*) Blind ant (*d*) All of these

74. Only bruchid which attacks groundnut in storage is

(*a*) *Callosobruchus maculates* (*b*) *Callosobruchus minutes*

(*c*) *Caryedon serratus* (*d*) None of these

75. Which of the following fly is the pest of pulses?

(*a*) Fruit fly (*b*) Stem fly

(*c*) White fly (*d*) Saw fly

76. Cut worm is a serious pest of

(*a*) Cotton (*b*) Citrus

(*c*) Gram (*d*) Arhar

77. Mustard sawfly is scientifically known as

(*a*) *Athalia lugens proxima* (*b*) *Earias vitella*

(*c*) *Virachola Isocrates* (*d*) *Pectinophora gossipiella*

78. The mustard sawfly lays eggs near

(*a*) Leaf margin (*b*) Roots

(*c*) Stem (*d*) Flower

79. Mustard sawfly pupates in

(*a*) Soil (*b*) Leaves

(*c*) Stem (*d*) None of these

80. Curly leaf tip with poor pod set in mustard is characteristic symptom of

(*a*) Aphid (*b*) Cutworm

(*c*) Leafhopper (*d*) Saw fly

81. Mustard saw fly belongs to the order

(*a*) Hemiptera (*b*) Diptera

(*c*) Hymenoptera (*d*) Lepidoptera

82. Mustard saw fly is

(*a*) Monophagy (*b*) Oligophagy

(*c*) Polyphagy (*d*) Omnivorous

83. *Dasineura lini* is a pest of

(*a*) Seasmum (*b*) Linseed

(*c*) Castor (*d*) Groundnut

84. Spotted bollworm is a pest of

(*a*) Citrus (*b*) Cotton

(*c*) Brinjal (*d*) Tomato

85. *Earias vitellla* is commonly known as

(*a*) Pink boll worm (*b*) Spotted boll worm

(*c*) Cotton white fly (*d*) Aphid

86. *Spilosoma obliqua* is commonly known as

(*a*) Red hairy caterpillar (*b*) Bihar hairy caterpillar

(*c*) Bud worm (*d*) Pink boll worm

87. Curling and upward yellowing of cotton is caused by

(*a*) Cotton boll worm (*b*) Cotton leaf hopper

(*c*) Cotton aphid (*d*) Red cotton bug

88. Pink bollworm is scientifically known as

(*a*) *Earias vitella* (*b*) *Pectinophora gossypiella*

(*c*) *Helicoverpa armigera* (*d*) None of these

89. Damaging stage of cotton jassid are

(*a*) Maggot (*b*) Adult

(*c*) Nymph (*d*) Both Nymph and Adult

90. Important pest of Mesta

(*a*) Aphid (*b*) Spiral borer

(*c*) Bihar Hairy Caterpillar (*d*) Capsule borer

91. Bihar hairy caterpillar belongs to the family

(*a*) Acrididae (*b*) Noctuidae

(*c*) Pyralidae (*d*) Apidae

92. Serious sucking pest of cotton is

(*a*) Pink boll worm (*b*) Leaf hopper

(*c*) Aphid (*d*) Thrips

93. Circular bore hole in the boll plugged by the head of a larva is characteristic symptom of

(*a*) Pink boll worm (*b*) Spotted boll worm

(*c*) American boll worm (*d*) Spiny boll worm

94. Flared squares in cotton is caused by

(*a*) *Spodoptera litura* (*b*) *Helicoverpa armigera*

(*c*) *Pectinophora gossypiella* (*d*) *Earias vitella*

95. The bollworm which covers the opening once it enters into the boll is

(*a*) Pink boll worm (*b*) Spotted boll worm

(*c*) American boll worm (*d*) Spiny boll worm

96. In cotton, gall like swelling near the collar region is caused by

(*a*) Stem weevil (*b*) Boll worm

(*c*) Red cotton bug (*d*) White fly

97. Curling of leaf upwards and yellowing of terminal cotton shoots is due to

(*a*) Whitefly (*b*) Cotton leaf hopper

(*c*) Bollworms (*d*) Cotton aphid

98. Marginal downward rolling, cupping, yellowing, browning, bronzing and drying of cotton leaves is characteristic symptom of

(*a*) Whitefly (*b*) Cotton leaf hopper

(*c*) Bollworms (*d*) Red cotton bug

99. Larvae of pink bollworm overwinter in

(*a*) Seed (*b*) Stem

(*c*) Soil (*d*) Leaf

100. The bacteria responsible for staining of the cotton fiber is

(a) *Bacillus gossypii*
(b) *Azotobactor* sp.
(c) *Nematosposa gossypii*
(d) None of these

101. Rosette flowers in cotton is caused by

(a) Pink boll worm
(b) Spotted boll worm
(c) Red cotton bug
(d) Spiny boll worm

102. Staining of lint in cotton is caused by

(a) Pink boll worm
(b) Spotted boll worm
(c) Red cotton bug
(d) Spiny boll worm

103. Colour of eggs of spotted bollworm is

(a) Black
(b) Sky blue
(c) Pink
(d) Green

104. Pink bollworm is a notorious pest of

(a) Wheat
(b) Cotton
(c) Sugarcane
(d) Pulses

105. Red cotton bug lay eggs on/in the

(a) Leaf
(b) Stem
(c) Soil
(d) Boll

106. Recently the outbreak of woolly aphid was on which crops

(a) Maize
(b) Okra
(c) Sugarcane
(d) Litchi

107. Sugarcane wooly aphid is scientifically known as

(a) *Eriosoma lanigera*
(b) *Cerato vacuno*
(c) *Lipaphis* spp.
(d) None of these

108. Bushy appearance with dead heart in sugarcane is damage symptoms of

(a) Root borer
(b) Top borer
(c) Stem borer
(d) Woolly aphid

109. Sugarcane root borer is scientifically known as

(a) *Earias* spp.
(b) *Emmelocera depresella*
(c) *Pectinophora* spp.
(d) *Helicoverpa armigera*

110. Pyrilla belongs to the family

(a) Jassidae
(b) Fulgoridae
(c) Coccidae
(d) Pyrrhocoridae

111. Sugarcane borer which does not belong to family pyralidae

(a) *Emmalocera depressella* (b) *Chilo infuscatelus*

(c) *Sesamia inferens* (d) *Chilo auricilia*

112. Bunchy top appearance in grown up sugarcane due to formation of side shoots is a characteristic symptom of

(a) Top shoot borer (b) Early shoot borer

(c) Inter node borer (d) All of these

113. Sugarcane setts filled with soil is characteristic symptom of

(a) Root borer (b) Scale insect

(c) Termites (d) Wooly aphid

114. Sugarcane root borer, *Emmalocera depresella* attacks

(a) Roots (b) Shoot

(c) Leaves (d) Flowers

115. Stripping of sugarcane leaves minimize the attack of

(a) Mealy bug (b) Woolly aphid

(c) Top borer (d) None of these

116. Shriveling and drying of sugarcanes with reduced internodes is caused by

(a) Root borer (b) Scale insect

(c) Termites (d) Wooly aphid

117. Anar butterfly is scientifically known as

(a) *Helicoverpa armigera* (b) *Earias vitella*

(c) *Virachola isocrates* (d) *Pectinophora gossipiella*

118. Fruit fly pupate in

(a) Soil (b) Fruit

(c) Leaf (d) Stem

119. Banana aphid is scientifically known as

(a) *Aphis craccivora* (b) *Thrips tabaci*

(c) *Pentalonia nigronervosa* (d) *Lipaphis ericimy*

120. Crown topping and gummosis of coconut is the damage symptoms of

(a) Red palm weevil (b) Coconut beetle

(c) Aphid (d) Mite

121. Grease painting is effective control of

(a) Mealy bug (b) Stem borer

(c) Aphid (d) Stone weevil

122. Mango mealy bug is scientifically known as

(a) *Apsilla* spp. (b) *Drosicha* spp.

(c) *Virachola* spp. (d) None of these

123. In Banana which type of nematode occurs

(a) *Tylenchulus* (b) *Trychodorus*

(c) *Rodophilus* (d) *Meloydogyne*

124. Amongst the following insects which one is monophagous

(a) *Spodoptera litura* (b) *Spilosoma oblicua*

(c) *Bombyx mori* (d) *Bemicia tabaci*

125. Bagging is recommended to control which of the following insect pests

(a) Anar butterfly (b) Lemon butterfly

(c) Mango stone weevil (d) Ber fruit fly

126. The caterpillars of *Indarbela quadrinotata* feed and damage

(a) Fruit (b) Flower

(c) Leaves (d) Bark

127. The larvae of fruit sucking moth feeds on

(a) Flower (b) Fruit

(c) Weed hosts (d) Roots

128. Rounded eggs are laid by which one among the following insects

(a) Castor butterfly (b) Lace wing fly

(c) Mosquito (d) Citrus butterfly

129. Single egg laying behavior is seen

(a) Castor butterfly (b) Sorghum shoot fly

(c) Citrus butterfly (d) All of these

130. Which of the following insect is monophagy?

(a) Yellow stem borer of rice (b) Mango nut weevil

(c) Coconut black headed caterpillar (d) All of these

131. Which of the following insect larvae looking like bird droppings?

(a) Slug caterpillar (b) Citrus butterfly

(c) House fly (d) Mantids

132. Fruit drooping at marble size stage is noticed in mango due to the infestation of

(a) Stem borer (b) Fruit fly

(c) Stone weevil (d) Mango mealy bug

133. Bore hole on pomegranate fruits surrounded by faecal material is characteristic symptom of

(*a*) Shoot borer (*b*) Leaf eater

(*c*) Anar butterfly or fruit borer (*d*) Bud borer

134. San Jose Scale passes the winter as

(*a*) Egg (*b*) Nymph

(*c*) Pupa (*d*) Adult

135. The insect pest of grapes which is not present in India is

(*a*) Stem Girdler (*b*) Aphid

(*c*) Grape Phyloxera (*d*) Leaf Cutter

136. Which of the following is a monophagous pest of mango?

(*a*) Red ant (*b*) Hopper

(*c*) Nut weevil (*d*) Stem borer

137. 'T' shaped marking on marble sized mango fruits is characteristic symptom of

(*a*) Red ant (*b*) Hopper

(*c*) Nut weevil (*d*) Stem borer

138. Drying and withering of terminal shoot in citrus with the presence of large number of ants is due to the attack of

(*a*) Citrus psyllid (*b*) Fruit sucking moth

(*c*) Fruit fly (*d*) Lemon butterfly

139. Shedding of citrus fruits with feeding punctures is a damage symptom of

(*a*) Fruit fly (*b*) Fruit sucking moth

(*c*) Fruit borer (*d*) Lemon butterfly

140. Cottony chusion scale is a pest of

(*a*) Cotton (*b*) Citrus

(*c*) Gram (*d*) Arhar

141. Sanjose scale is a pest of

(*a*) Almond and Apple (*b*) Pear and Plum

(*c*) Litchi and Mango (*d*) None of these

142. In apple, infested region on bark turns into pink colour and pink colour encrustation on apple fruit is a characteristic symptom of

(*a*) Wooly aphid (*b*) San Jose Scale

(*c*) Stem borer (*d*) None of these

143. Ashy coat on apple stem with galled roots is due to

(a) Wooly aphid (b) San Jose Scale

(c) Stem borer (d) None of these

144. Spiral cut on the grapevine stem is the damage caused by

(a) Stem borer (b) Stem girdler

(c) Flea beetle (d) Mealy bug

145. In beetle vine, yellowing of leaves with gray coloured armored scale covering them is due to

(a) Mites (b) Beetle Vine scale

(c) Thrips (d) Leafhoppers

146. Which one of the following pest is an introduced pest of coffee?

(a) Green bug (b) Berry borer

(c) Stem borer (d) Mealy bug

147. Geometrical cutting on coconut founds is due to

(a) Rhinoceros beetle (b) Red palm weevil

(c) Black headed caterpillar (d) Mite

148. Gummosis and crown topping in coconut is characteristic symptom of

(a) Rhinoceros beetle (b) Red palm weevil

(c) Black headed caterpillar (d) Mite

149. In coconut, stem/root injection technique is used for management of the

(a) Rhinoceros beetle (b) Red palm weevil

(c) Mites (d) Black headed caterpillar

150. Which of the following insect cause damage to the trunk of Mango?

(a) Mealy bug (b) Stone weevil

(c) Stem borer (d) Fruit fly

151. *Cosmopolites sordidus* related with which of the following crop

(a) Mango (b) Banana

(c) Grape (d) Guava

152. The pest that cause annoys in climbing tree

(a) Mango stone weevil (b) Mango hopper

(c) Mango mealy bug (d) Fruit fly

153. *Virachola isocrates* is a severe pests of

(a) Pomegranate (b) Mango

(c) Apple (d) Citrus

154. Banana rhizome weevil is belongs to the family

(*a*) Dermestidae (*b*) Chrysomilidae

(*c*) Bostrichidae (*d*) Curculienidae

155. Banana rhizome weevil is scientifically known as

(*a*) *Eriosoma lanigera* (*b*) *Cerato vacuno*

(*c*) *Aphis gossypi* (*d*) *Cosmopolites sordidus*

156. *Henosepilachna dodecastigma* is a pest of

(*a*) Potato (*b*) Tomato

(*c*) Brinjal (*d*) All of these

157. Occurrence of half moon shaped notching on brinjal leaves is due to

(*a*) *Myllocerus subfasciantus* (*b*) *Leucinodes orbonalis*

(*c*) *Earias vitella* (*d*) *Aphis craccivora*

158. The red pumpkin beetle belongs to the family

(*a*) Chrysomelidae (*b*) Coccidae

(*c*) Dermastidae (*d*) Coccinelidae

159. Spiraling whitefly is scientifically known as

(*a*) *Bemesia tabaci* (*b*) *Dialeurodes* spp.

(*c*) *Aleurodidus disperses* (*d*) *Aleurolobus* spp

160. Name of potato pest attack both in field and storage is

(*a*) Cut worm (*b*) Root borer

(*c*) Potato tuber moth (*d*) Tobacco caterpillar

161. Osmeteria presents in

(*a*) Cabbage butterfly (*b*) Lemon butterfly

(*c*) Aphid (*d*) White grub

162. Most important pest of Brinjal is

(*a*) Aphid (*b*) Stem fly

(*c*) Gall fly (*d*) Fruit and Shoot borer

163. Which of the following bug is a pest of brinjal?

(*a*) Painted bug (*b*) Plant bug

(*c*) Stink bug (*d*) Lace bug

164. Mines on leaves and bore holes on tender shoot and tubers in potato is caused by

(*a*) Cutworm (*b*) Potato tuber moth

(*c*) Aphid (*d*) Beetle

165. Which pest of potato attacks the tuber both in field and storage?

(a) Cutworm (b) Potato tuber moth

(c) Potato aphid (d) Beetle

166. Notching of brinjal leaves is caused by feeding of

(a) Stem borer (b) Fruit borer

(c) Ash weevil (d) Hadda beetle

167. Mining and skeletonization of cabbage leaves is caused by

(a) Diamond back moth (b) Leaf webber

(c) Head borer (d) Semilooper

168. Release of larval parasitoid, *Cotesia plutella* is to reduce the damage of the following pest

(a) Diamond back moth (b) Leaf webber

(c) Head borer (d) Semilooper

169. Irregular holes on cucurbit leaves is the damage symptom of

(a) Fruit flies (b) Flea beetle

(c) Red pumpkin beetle (d) None of these

170. Upwardly curling of chilli leaves with rat tail like tip is due to the attack by

(a) Thrips (b) Mites

(c) Scales (d) Leafhoppers

171. Which one of the following is a serious pest of onion and garlic?

(a) Thrips (b) Mealy bug

(c) Leafhopper (d) Bulb borer

172. Bore hole on the pseudostem, berries with accumulation of fecal pellets in cardamom is due to the attack of

(a) Mealy bug (b) Stem borer

(c) Root borer (d) Capsule borer

173. In pepper, small holes in leaves and berries of pepper is due infestation of

(a) Flea beetle (b) Shot hole borer

(c) Pollue beetle (d) Semilooper

174. Pollu beetle is a major pest of

(a) Cardamom (b) Beetle vine

(c) Black pepper (d) All of these

175. Tip drying in onion is due to the infestation of

(*a*) Shoot borer
(*b*) Thrips
(*c*) Aphid
(*d*) Bulb borer

176. Leaf curls down, inverted boated shaped chilli leaf with elongated petiole is due to the attack of

(*a*) Thrips
(*b*) Aphids
(*c*) Jassids
(*d*) Mites

177. The insect pest of potato which gained entry into India is

(*a*) Cut worm
(*b*) Epilachna beetle
(*c*) Potato Tuber Moth
(*d*) Stem Borer

178. Which insects (as immature) do not damage horticultural crops?

(*a*) Sawflies
(*b*) Butterflies
(*c*) Whiteflies
(*d*) Caddisflies

179. Which of the following insect is a gregarious pest?

(*a*) Sorghum shoot fly
(*b*) Spotted boll worm
(*c*) Migratory locust
(*d*) Rice case worm

180. Example of oligophagous pest is

(*a*) Mango stone weevil
(*b*) Diamond back moth
(*c*) Grasshopper
(*d*) Yellow Stem Borer

181. Family of Hymenoptera having crop pests

(*a*) Apidae
(*b*) Tenthredinidae
(*c*) Braconidae
(*d*) Bombidae

182. Most destructive polyphagous insect is

(*a*) Whitefly
(*b*) Locust
(*c*) Mealy bugs
(*d*) Stem borer

183. Preferred feeding zone of Jute Semi-looper, *Anomis sabulifera* is

(*a*) Bottom older leaves
(*b*) Top 7 - 9 leaves
(*c*) Whole plant
(*d*) Two middle leaves

184. Skeletonization of tobacco leaves is caused by

(*a*) Tobacco caterpillar
(*b*) Gram cutworm
(*c*) Leafhopper
(*d*) Leaf miner

185. Cannibalism is seen in

(*a*) Gram pod borer
(*b*) Tobacco caterpillar
(*c*) Red gram plume moth
(*d*) All of these

186. White grubs lay eggs

(a) At floral parts
(b) Underside of leaf
(c) On plant debris
(d) Singly in loose soil

187. The insect laying eggs in silken stalks of about 1.25 mm length to prevent cannibalism among the larvae of

(a) Lace wing bug
(b) Lace wing fly
(c) Diamond Backed Moth
(d) White fly

188. Which of the following is true?

(a) Compact clay soil favour termite
(b) Red soil favour termite
(c) Sandy soil favour white grub
(d) Both (b) & (c)

189. The incidence of "White Grubs" is heavy in

(a) Allways heavily water looged fields
(b) Dry fields having sandy loam soil
(c) Dry fields of heavy clay soil
(d) Desert land having little vegetation

190. The largest locust swarm recorded so far

(a) 100 square kilometers
(b) 200 square kilometers
(c) 300 square kilometers
(d) 400 square kilometers

191. Gram pod borer feeds on which of following crops

(a) Gram only
(b) Rice
(c) Tomato only
(d) Variety of Crops

192. Red hairy caterpillar pupate in

(a) Fallen leaves
(b) Soil
(c) Flower bud
(d) None of these

193. Which animal is known as natural filler of soil?

(a) Insects
(b) Earthworms
(c) Mite
(d) Nematode

194. Insects feeds on plants of several genera within family known as

(a) Monophagous
(b) Oligophagous
(c) Polyphagous
(d) Stenophagous

195. Generally locust eat almost all vegetation except

(a) *Datura stramonium*
(b) *Melia azadirach*
(c) *Calotropis procera*
(d) All of these

196. Example of pest which has got theylitokous parthenogenesis in its life cycle

(*a*) Aphid (*b*) Grasshopper
(*c*) Thrips (*d*) Whitefly

197. The average locust swarm spreads over

(*a*) 5 square kilometers (*b*) 10 square kilometers
(*c*) 15 square kilometers (*d*) 20 square kilometers

198. Eggs are looking like miniature batteries of

(*a*) Cockroach (*b*) House fly
(*c*) Stink bug (*d*) Yellow stem borer

199. Locusts lay eggs

(*a*) On soil surface (*b*) In the water
(*c*) In the soil (*d*) On the tree

200. *Odontotermus obesus* commonly known as

(*a*) White fly (*b*) Thrips
(*c*) Termite (*d*) None of these

201. Most notorious polyphagous pest is

(*a*) *Helicoverpa armigera* (*b*) *Agrotis* spp.
(*c*) *Chilo partellus* (*d*) None of these

202. House of termite is called

(*a*) Cave (*b*) Mound
(*c*) Hive (*d*) Tunnel

203. Mole rat is scientifically known as

(*a*) *Ratus ratus* (*b*) *Bandicota bengalensis*
(*c*) *Mus musculus* (*d*) None of these

204. Example of cosmopolitan polyphagous insect is

(*a*) Mango stone weevil (*b*) Oriental fruit fly
(*c*) Locust (*d*) Bihar hairy caterpillar

205. Generally nematicide is used with

(*a*) Top dressing (*b*) Fertilizer
(*c*) Irrigation water (*d*) None of these

206. Plant parasitic nematode generally attacks on

(*a*) Roots (*b*) Leaf
(*c*) Stem (*d*) Fruit

207. Golden cyst nematode attack on

(a) Wheat
(b) Rice
(c) Gram
(d) Cotton

208. 1st reported plant parasitic nematode is

(a) *Angunia tritici*
(b) *Heterodera* spp.
(c) *Aphalenchoides* spp.
(d) *Tylencorinchus* spp.

209. Field rat is scientifically known as

(a) *Ratus meltoda*
(b) *Mus musculus*
(c) *Ratus ratus*
(d) None of these

210. The feeding site of root knot nematode is

(a) Xylem
(b) Phloem
(c) Vascular bundle
(d) None of these

211. Molya disease of wheat was discovered by

(a) Vasudeva
(b) Van
(c) Taylor
(d) Dastur

212. Causal organism of tungro disease of rice is

(a) Bacteria
(b) Nematode
(c) Virus
(d) Fungi

213. Causal organism of ufra disease of rice is

(a) *Anguinia* spp.
(b) *Dietylenchus angustus*
(c) *Heterodera rostreii*
(d) *Longidora* sp.

214. Nematode was first reported by

(a) Ishida
(b) Needham
(c) Iswariya
(d) Pruthi

215. In Nilgiri hills which nematode is most hazardous to potato

(a) Cyst nematode
(b) Golden nematode
(c) Gall nematode
(d) Root knot nematode

216. Which of the following crop mostly damaged by *Meloidogyne incognita*?

(a) Maize
(b) Wheat
(c) Tomato
(d) Cotton

217. The grub of *Holotrichia consanguinea* emerge during

(a) Morning
(b) Day time
(c) Night
(d) None of these

218. Which of the following is a polyphagous pest?

(*a*) Rice Gundhi bug (*b*) Gram Pod borer

(*c*) Sugarcane top borer (*d*) Red Pumpkin beetle

219. Which crop is most susceptible crop for root knot nematode?

(*a*) Cotton (*b*) Wheat

(*c*) Brinjal (*d*) Okra

220. Which of the following is the primary source of Wheat Ear-cockle inoculums?

(*a*) Soil (*b*) Air

(*c*) Infected seed (*d*) All of these

221. Nematodes are generally found in the soil at a depth of

(*a*) 6 - 9 mm (*b*) 6 - 9 cm

(*c*) 6 - 9 inch (*d*) 9 - 12 inch

222. Which of the following is effective stage of Nematode?

(*a*) 1st stage (*b*) 2nd stage

(*c*) 3rd stage (*d*) Adult stage

223. The most important characteristic of nematode is

(*a*) Cyst forming in nature (*b*) Acellular

(*c*) Endo-parasite in nature (*d*) Ecto-parasite in nature

224. Dagger nematode is also known as

(*a*) *Xiphinema* (*b*) *Longidorus*

(*c*) *Trichodorus* (*d*) *Anguina*

225. Citrus Nematode is

(*a*) *Dictylenchus* (*b*) *Tylenculus*

(*c*) *Longidorus* (*d*) *Heterodora*

226. Vermiculture technology is associated with

(*a*) Silkworm (*b*) Cutworm

(*c*) Earthworm (*d*) Armyworm

227. Which of the following pests is migratory in nature?

(*a*) Aphids (*b*) Locust

(*c*) Danaus (*d*) All of these

228. Ballooning is behavioral character of

(*a*) Gypsy moth (*b*) Mealy bug

(*c*) Aphid (*d*) Bark borer

Fill in the Blanks

1. The female moth of yellow stem borer has bright yellowish brown fore wings with a__________spot on each.
2. The larva of swarming caterpillar, *Spodoptera mauritia* pupated in an__________in soil.
3. Rice thrips lacerate the tender leaves and suck up the plant sap causing__________of leaves.
4. Infestation of rice mealy bug causes sunken oval or round patches in a normal crop and the plants may either die or put forth distorted panicles with__________grains.
5. Both nymphs and adults suck the plant sap and cause__________of leaves.
6. Long winged forms of brown plant hoppers are also known as__________.
7. Short winged forms of brown plant hoppers are also known as__________.
8. The grub of rice hispa feeds by__________into leaves.
9. The adult of rice hispa feeds by__________the green matter of leaves.
10. The larva of coconut black headed caterpillar feeds by__________the green matter of leaves.
11. The larva of coconut black headed caterpillar pupate within__________.
12. In case of severe attack of coconut black headed caterpillar due to large scale drying of leaflets the plantation presents a__________appearance from a distance.
13. Hopper burn symptoms of cotton is caused by__________.
14. Jute stem girdler, *Nupserha bicolor postbrunnea* girdles the stem at two levels before it starts__________.
15. The larvae of jute semilooper, *Anomis sabulifera* feed at__________canopy.
16. The grub of jute weevil, *Apion corchori* makes tunnel into__________.
17. Gall like swelling on jute stem is formed due to damage of__________.
18. Serpentine leaf miner, *Liriomyza trifolii* is belongs to the family__________.
19. The sweet potato weevil, *Cylas formicarius* is an important pest both at__________and store.

20. Among the mango hoppers the largest one is_________.

21. Among the mango hoppers the smallest one is_________.

22. Presence of winding galleries of frassy web on the stem of citrus indicate the infestation of_________.

23. Citrus leaf miner, *Phyllocnistis citrella* belongs to the family_________.

24. Larvae of citrus leaf miner, *Phyllocnistis citrella* is apodous and pupate inside the tunnel in_________.

25. Wilting of coffee branches occurs due to the infestation of_________.

Match the Following

1. Match the following insects regarding the damage symptom of grapes

(a)	Shot holes on terminal leaves of grapes	i)	Stem girdler
(b)	Corky out growth on grapevine berries	ii)	Mealy bug
(c)	Sticky, shriveled grapevine berries with sooty mould	iii)	Berry thrips
(d)	Spiral cut on the grapevine stem	iv)	Grapevine flea beetle

2. Matched the insects of citrus with their damage symptom

(a)	Galleries are found inside the stem	i)	Citrus butterfly
(b)	Both nymph and adult suck the cell sap from leaf	ii)	Citrus psylla
(c)	Defoliation in citrus	iii)	Citrus shoot and bark borer
(d)	Zigzag leaf mines in citrus leaves	iv)	Citrus leaf miner

3. matched the pest of mango with their damage symptom

(a)	The young fruits become juice less	i)	Red ant
(b)	Webbing mango leaves with petiole intact	ii)	Mealy bugs
(c)	Drying of inflorescence and shedding of flowers	iii)	Mango hopper
(d)	Bore hole on mango stem with protrusion of frass	iv)	Stem borer

4. Match the following insects with their damage symptoms

(a)	Corky out growth or scab formation	i)	Tea mosquito bug
(b)	Yellowing of leaves and development of sooty mould	ii)	Fruit fly

(c)	Leaves fall off and twigs dry up	iii)	Scale insect
(d)	Maggots destroy the fruit pulp	iv)	Spiraling white fly

5. **Match the following insects with their damage symptoms on vegetable**

(a)	Bore holes on bhendi fruits	i)	Hadda beetle
(b)	Mining and skeletonization of brinjal	ii)	Sweet potato weevil
(c)	Bore holes on sweet potato tuber both in field and storage	iii)	Fruit borer
(d)	Notching of brinjal leaves	iv)	Ash weevil

6. **Match the following insects with their damage symptoms on coconut**

(a)	Dried leaflets with gallery in older coconut fronds	i)	Lacewing bug
(b)	Yellowing of coconut leaves in patches	ii)	Black headed caterpillar
(c)	White speckling on upper surface of coconut leaves	iii)	Rhinoceros beetle
(d)	Shot holes on fronds of coconut trunk	iv)	Scale

7. **Match the following insects with their damage symptoms on sugarcane**

(a)	Daead hearts	i)	Termite
(b)	Bore holes on internodes	ii)	Early shoot borer
(c)	Bunchy top	iii)	Inter node borer
(d)	Sugarcane setts filled with soil	iv)	Top borer

8. **Match the following insects with their damage symptoms on cotton**

(a)	Drooping shoots and flared squares in cotton	i)	Spotted bollworm
(b)	Rosette bloom in cotton	ii)	Pink bollworm
(c)	Gall like swelling near the collar region of cotton	iii)	Stem weevil
(d)	Marginal downward rolling, cupping and yellowing of cotton leaves	iv)	Leaf hopper

9. **Match the following insects with their damage symptoms on groundnut**

(a)	Total defoliation in groundnut	i)	Red hairy caterpillar
(b)	Webbing and drying of terminal leaflets in groundnut	ii)	*Spodoptera litura*
(c)	Sucking cell sap from under side of leaves	iii)	Groundnut leaf miner
(d)	Wilting of groundnut plants in patches with damaged roots	iv)	White grub

10. Match the following insects with their damage symptoms on pulse

(a) Circular bore holes	i) Gram pod borer
(b) Irregular bore hole	ii) Plumemoth
(c) Pin head sized bore hole	iii) Blister beetle
(d) Flowers partially eaten	iv) Podfly

11. Match the following insects with their damage symptoms on rice

(a) Infected leaves turn brown	i) Gundhi bug
(b) Buggy odour in rice during milky stage	ii) Green Leaf Hopper
(c) Nibbling of rice ear head	iii) Case worm
(d) Cut leaf bit floating in water	iv) Grass hopper

12. Match the following insects with their damage symptoms on sorghum

(a) Dead heart, reddening of Sorghum midrib	i) Sorghum stem borer
(b) Pupa case producing in between glumes	ii) Sorghum midge
(c) Reddening of lower leaves with silken web	iii) Sorghum mite
(d) Chaffyness	iv) Ear head bug

13. Match the following insects with their damage symptoms on pulse

(a) Flowers partially eaten	i) Blister beetle
(b) Pin head sized bore hole	ii) Spotted pod borer
(c) Webbing of flower inflorescence	iii) Pod fly
(d) Shriveled seeds inside the pulse pods	iv) Pod bug

14. Match the insect pests and their host plant

(a) *Virachola Isocrates*	i) Tea
(b) *Callosobruchus chinensis*	ii) Green gram
(c) *Dasineura lini*	iii) Linseed
(d) *Helopeltis theivora*	iv) Pomegranate

15. Match the insect with their scientific name

(a) *Hypothenemus hampei*	i) Coffee berry borer
(b) *Eriosoma lanigerum*	ii) Spiraling whitefly
(c) *Pentalonia nigronervosa*	iii) Banana aphid
(d) *Aleurodicus dispersus*	iv) Apple woolly aphid

16. Match the insect pests with their family

(a)	White backed plant hopper	i)	Delphacidae
(b)	Yellow Stem Borer	ii)	Pyralidae
(c)	Diamond Backed Moth	iii)	*Yponomeutidae*
(d)	Red pumpkin beetle	iv)	Chrysomelidae

17. Match the insect pests with their scientific names

(a)	Citrus leaf miner	i)	*Phyllocnistis citrella*
(b)	Citrus psylla	ii)	*Diaphorina citri*
(c)	Citrus black fly	iii)	*Aleurocanthus woglumi*
(d)	Fruit piercing moth	iv)	*Ophideres conjuncta*

18. Match the common name of aphid with their scientific name

(a)	*Brevicoryne brassicae*	i)	Green peach aphid
(b)	*Myzus persicae*	ii)	Cowpea aphid
(c)	*Aphis craccivora*	iii)	Turnip aphid
(d)	*Lipaphis erysimi*	iv)	Cabbage aphid

19. Match the insects with their methods of management

(a)	Aphid	i)	Extraction with hooks
(b)	White grub	ii)	Yellow sticky trap
(c)	American bollworm	iii)	Growing of Marigold
(d)	Rhinocerus beetle	iv)	Deep summer ploughing

Answers

Multiple Choice Questions

1	(a)	21	(b)	41	(a)	61	(a)
2	(c)	22	(c)	42	(b)	62	(d)
3	(b)	23	(b)	43	(b)	63	(b)
4	(d)	24	(a)	44	(a)	64	(d)
5	(a)	25	(c)	45	(b)	65	(c)
6	(a)	26	(b)	46	(a)	66	(c)
7	(d)	27	(b)	47	(a)	67	(c)
8	(b)	28	(a)	48	(d)	68	(a)
9	(c)	29	(a)	49	(c)	69	(b)
10	(d)	30	(d)	50	(c)	70	(a)
11	(b)	31	(a)	51	(b)	71	(c)
12	(d)	32	(c)	52	(b)	72	(b)
13	(a)	33	(a)	53	(a)	73	(d)
14	(c)	34	(d)	54	(b)	74	(c)
15	(d)	35	(a)	55	(b)	75	(b)
16	(a)	36	(d)	56	(b)	76	(c)
17	(b)	37	(b)	57	(b)	77	(a)
18	(a)	38	(a)	58	(d)	78	(a)
19	(a)	39	(a)	59	(a)	79	(a)
20	(a)	40	(b)	60	(b)	80	(a)

81	(c)	115	(a)	149	(b)	183	(b)
82	(b)	116	(b)	150	(c)	184	(a)
83	(b)	117	(c)	151	(b)	185	(d)
84	(b)	118	(a)	152	(b)	186	(d)
85	(b)	119	(c)	153	(a)	187	(b)
86	(b)	120	(a)	154	(d)	188	(d)
87	(b)	121	(a)	155	(d)	189	(b)
88	(b)	122	(b)	156	(d)	190	(d)
89	(d)	123	(c)	157	(a)	191	(d)
90	(b)	124	(c)	158	(a)	192	(b)
91	(b)	125	(a)	159	(c)	193	(b)
92	(b)	126	(d)	160	(c)	194	(b)
93	(c)	127	(c)	161	(a)	195	(d)
94	(d)	128	(d)	162	(d)	196	(a)
95	(a)	129	(d)	163	(d)	197	(b)
96	(a)	130	(d)	164	(b)	198	(c)
97	(b)	131	(b)	165	(b)	199	(c)
98	(b)	132	(c)	166	(c)	200	(c)
99	(a)	133	(c)	167	(a)	201	(a)
100	(c)	134	(b)	168	(a)	202	(b)
101	(a)	135	(c)	169	(c)	203	(b)
102	(c)	136	(c)	170	(a)	204	(c)
103	(b)	137	(c)	171	(a)	205	(c)
104	(b)	138	(a)	172	(d)	206	(a)
105	(c)	139	(b)	173	(c)	207	(a)
106	(c)	140	(b)	174	(c)	208	(a)
107	(b)	141	(a)	175	(b)	209	(b)
108	(b)	142	(b)	176	(d)	210	(c)
109	(b)	143	(a)	177	(c)	211	(a)
110	(b)	144	(b)	178	(d)	212	(c)
111	(c)	145	(b)	179	(c)	213	(b)
112	(a)	146	(b)	180	(b)	214	(b)
113	(c)	147	(a)	181	(b)	215	(b)
114	(b)	148	(b)	182	(b)	216	(c)

217	(c)	220	(c)	223	(c)	226	(c)
218	(b)	221	(b)	224	(a)	227	(d)
219	(d)	222	(b)	225	(b)	228	(a)

Fill in the Blanks

1. Black
2. Earthen cocoon
3. Rolling
4. Chaffy
5. Browning
6. Macropterous
7. Brachypterous
8. Mine
9. Scraping
10. Scraping
11. Silken galleries
12. Burnt up
13. *Amrasca biguttula biguttula*
14. Oviposition
15. Upper
16. Pith
17. *Apion corchori*
18. Agromyzidae
19. Field
20. *Amritodus atkinsoni*
21. *Idioscopus clypealis*
22. *Indarbela tetraonis*
23. Gracillariidae
24. Silken web
25. Coffee borer

Match the Following

1. a) - iv, b) - iii, c) - ii, d) - i
2. a) - iii, b) - ii, c) - i, d) - iv
3. a) - ii, b) - i, c) - iii, d) - iv
4. a) - i, b) - iv, c) - iii, d) - ii
5. a) - iii, b) - i, c) - ii, d) - iv
6. a) - ii, b) - iv, c) - i, d) - iii
7. a) - ii, b) - iii, c) - iv, d) - i
8. a) - i, b) - ii, c) - iii, d) - iv
9. a) - i, b) - iii, c) - ii, d) - iv
10. a) - i, b) - ii, c) - iv, d) - iii
11. a) - ii, b) - i, c) - iv, d) - iii
12. a) - i, b) - ii, c) - iii, d) - iv
13. a) - i, b) - iii, c) - ii, d) - iv
14. a) - iv, b) - ii, c) - iii, d) - i
15. a) - i, b) - iv, c) - iii, d) - ii
16. a) - i, b) - ii, c) - iii, d) - iv
17. a) - i, b) - ii, c) - iv, d) - iii
18. a) - iv, b) - i, c) - ii, d) - iii
19. a) - ii, b) - iv, c) - iii, d) - i

Chapter 12

Pests of Stored Products and their Management

Multiple Choice Questions

1. **With high rate of multiplication within one season, the stored grain pests may destroy grains upto**

   (a) 7 to 10 per cent (b) 10 to 12 per cent

   (c) 12 to 15 per cent (d) 17 to 20 per cent

2. **Pre harvesting attack mainly lead to quantity only, but attack during post harvest lead to**

   (a) Quantitative damage (b) Qualitative damages

   (c) Both *(a)* & *(b)* (d) None of these

3. **Insect pests of stored products attacks whole grains called as**

   (a) Primary pests (b) Secondary pests

   (c) Key pests (d) None of these

4. **Insect pests attacks broken grains are known as**

   (a) Primary pests (b) Secondary pests

   (c) Key pests (d) None of these

5. **Which of the following is true for Methyl bromide fumigation?**

   (a) All insect stages are effective and requires less exposure period

   (b) It is widely used for plant quarantine and food processing facilities

   (c) It destroys stratospheric ozone layer

   (d) All of these

6. **Stored products insect pests attacks the matured grains at the field called**
   (a) Contamination (b) Cross infestation
   (c) Hidden infestation (d) None of these

7. **When the fresh stocks of grain infected from old infected stocks known as**
   (a) Contamination (b) Cross infestation
   (c) Hidden infestation (d) None of these

8. **Cross infestation may occur during**
   (a) Threshing (b) Transportation
   (c) Both (a) & (b) (d) None of these

9. **The *Acarcus siro* attacks stored grains is a/an**
   (a) Insect (b) Mite
   (c) Fungus (d) Rodent

10. **Mite attacks which portion of the grains**
    (a) Whole grain (b) Germ portion
    (c) Husk (d) All of these

11. **The fumigant used for the management of stored grain pests**
    (a) Methyl bromide (b) Sulfuryl floride (Profume)
    (c) Phosphine (d) All of these

12. **Which of the following is true for Phosphine fumigation?**
    (a) Larval and adult stages are more susceptible than egg and pupae
    (b) It has little reaction with most commodities
    (c) It does not affect the germination and milling qualities
    (d) All of these

13. **Rice weevil, pulse beetle, lesser grain borer, cigarette beetle, sweet potato weevil and potato tuber moth of store products are**
    (a) Internal feeders (b) External feeders
    (c) Scavengers (d) Soil builder

14. **Cigarette beetle, *Lasioderma serricorne* belongs to the familly**
    (a) Anobiidae (b) Apionidae
    (c) Curculinidae (d) None of these

15. **The dropping are spindle shaped in**
    (a) Roof rat (b) Norway rat
    (c) Mouse (d) None of these

16. Sweet potato weevil, *Cylas formicarius* is

(*a*) Anobiidae (*b*) Apionidae

(*c*) Curculinidae (*d*) None of these

17. Red rust flour beetle, khapra beetle, Indian meal moth, almond moth and rice moth are

(*a*) Internal feeders (*b*) External feeders

(*c*) Scavengers (*d*) Soil builder

18. The adult of *Sitophilus oryzae* fly to the fields in the vicinity to infest the grains cause

(*a*) Hidden infestation (*b*) Cross infestation

(*c*) Contamination (*d*) All of these

19. Which stage of *Sitophilus* attacks rice grain?

(*a*) Adult (*b*) Larva

(*c*) Pupa (*d*) Both adult and larvae

20. *Sitophilus oryzae* unable to breed when the grain moisture content is less than

(*a*) 09 per cent (*b*) 10 per cent

(*c*) 12 per cent (*d*) 13 per cent

21. The moth is buff-coloured with pointed hind wing bearing a heavy fringe of hairs

(*a*) *Sitotroga cerealella* (*b*) *Corcyra cephalonica*

(*c*) *Cadra cautella* (*d*) None of these

22. The infesting and hibernating stage of *Sitotroga cerealella* is

(*a*) Adult (*b*) Larva

(*c*) Both (*a*) & (*b*) (*d*) None of these

23. Indian meal moth (*Plodia interpunctella*) larvae feed on grains covered under

(*a*) A silken web (*b*) Straw

(*c*) Chaffe (*d*) All of these

24. One Aluminium Phosphide tablet (three grams) releases__________grams of Phosphine gas.

(*a*) One (*b*) Two

(*c*) Three (*d*) Four

25. Hidden infestation can be detected by

(*a*) Staining method (*b*) Ninhydrin method

(*c*) KOH method (*d*) All of these

26. *Corcyra cephalonica* adults are

(*a*) Nocturnal (*b*) Crepuscular

(*c*) Diurnal (*d*) Matinal

27. The larvae of *Corcyra* feeds the grains under silken web like shelters, preferring the

(*a*) Whole grains (*b*) Partially damaged grains

(*c*) Grain dust (*d*) None of these

28. The almond moth (*Cadra cautella*) larva during feeding

(*a*) Spin silken tubes (*b*) Make silken cocoon

(*c*) Break the nut (*d*) None of these

29. The adults of khapra beetle (*Trogoderma granarium*) are

(*a*) A good flier (*b*) Cannot fly

(*c*) Can swim (*d*) A good buzzer

30. The larva of *Trogoderma* feed near the embryo and pupates within the last larval skin and hibernate

(*a*) Inside the grains (*b*) In cracks and crevices

(*c*) On grain dust (*d*) All of these

31. Upto 4 years without feeding the *Trogoderma granarium* is resistant to

(*a*) Temperature (*b*) Humidity

(*c*) Starvation (*d*) All of these

32. The larvae and adults of *Tribolium* unable to feed on

(*a*) Whole grains (*b*) Flour

(*c*) Milled grain (*d*) All of these

33. Lesser grain borer, *Rhizopertha dominica* originating in

(*a*) Indonesia (*b*) Australia

(*c*) India (*d*) Japan

34. The adult of *Rhizopertha dominica* is metalic (shining) dark brown or black in colour and head covered by hood shaped

(*a*) Metanotum (*b*) Pronotum

(*c*) Mesonotum (*d*) None of these

35. Which of the following is a Acute rodenticides (single dose and quick acting)?

(*a*) Zinc phosphide (*b*) Barium carbonate

(*c*) Both (*a*) & (*b*) (*d*) None of these

36. The hibernating stage of *Rhizopertha dominica* is

(*a*) Larva (*b*) Adult

(*c*) Pupa (*d*) Either (*a*) or (*b*)

37. Which of the following pests only grubs cause damage?

(*a*) *Caryedon serratus* (*b*) *Callosobruchus chinensis*

(*c*) *Trogoderma granarium* (*d*) All of these

38. Pusa bin is used for the control of

(*a*) *Sitophilus* (*b*) *Callosobruchus*

(*c*) *Trogoderma* (*d*) *Rhizopertha*

39. Which stage of *Rhizopertha dominica* are more harmful?

(*a*) Larva (*b*) Adult

(*c*) Pupa (*d*) Either (*a*) or (*b*)

40. The first inster larvae of *Rhizopertha dominica* are straight and can bore into grains other stages are curved and unable to bore and depends upon the flour produced by

(*a*) First inster larvae (*b*) Adult

(*c*) Other insect (*d*) All of these

41. In saw toothed grain beetle (*Oryzaephilus surinamensis*) the saw (6 + 6 = 12) are found on each side of

(*a*) Prothorax (*b*) Mesothorax

(*c*) Metathorax (*d*) 1st segment of abdomen

42. Cigarette beetle, *Lasioderma serricorne* larvae and adult attacks on tobacco products when sugar content of the product is

(*a*) Low (*b*) Moderate

(*c*) High (*d*) Nil

43. The larvae of *Callosobruchus* feeds within the grains and pupate generally at

(*a*) Inside the grain (*b*) Outside the grain

(*c*) Both (*a*) & (*b*) (*d*) None of these

44. Fungi attacks the stored products are

(*a*) *Aspergillus* and *Penicillum* (*b*) *Rhizopus* and *Fusarium*

(*c*) *Alternaria* and *Calvularia* (*d*) All of these

45. Which of the following refers to Zinc phosphide poison?

(*a*) Slow poison (*b*) Anticoagulant

(*c*) Acute poison (*d*) Mild poison

46. Fungi produce poisonous substances, which is toxic to vertebrates also is known as

(*a*) Mycotoxins (*b*) Myotoxins

(*c*) α toxin (*d*) α toxin

47. *Aspergillus flavus* produced which toxin in groundnut and groundnut meals

(*a*) Aflatoxin (*b*) α toxin

(*c*) α toxin (*d*) None of these

48. In store maize at high humidity and low temperature *Penicillium* produced grain caking and

(*a*) Blue eye (*b*) Black eye

(*c*) Red eye (*d*) Yellow eye

49. During storage, moisture content above 12 per cent and temperature of 30 - 32°C constitute a favourable environment for the development of

(*a*) Insect eggs (*b*) Fungal spores

(*c*) Pest population (*d*) None of these

50. The only stored grain pest of groundnut which attacks both kernels and pods

(*a*) *Corcyra cephalonica* (*b*) *Tribolium castaneum*

(*c*) *Caryedon serratus* (*d*) None of these

51. Insect start attacking the pods in field itself and infestation is carried to the godowns

(*a*) *Callosobruchus chinensis* (*b*) *Caryedon serratus*

(*c*) Both (*a*) & (*b*) (*d*) None of these

52. Due to movement and respiration of insect as well as grain respiration temperature of grain mass increase is known as

(*a*) Grain Burning (*b*) Hot Spot

(*c*) Grain Drying (*d*) None of these

53. The hot-humid air from "Hot Spots" go to upper cooler region and leave the moisture and droplets of water appear in upper surface of grain cause

(*a*) Grain Weeping (*b*) Grain Sweating

(*c*) Grain Burning (*d*) All of these

54. Due to "Grain Sweating" storage fungi become active and starts

(*a*) Grain breaking (*b*) Grain caking

(*c*) Grain dusting (*d*) None of these

55. Which of the following is a chronic rodenticides (Multi dose and slow acting)?

(*a*) Warfarin (*b*) Bromodiolone

(*c*) Both (*a*) & (*b*) (*d*) None of these

56. To protect the infestation of pulse beetle, cover the pulses stored in bulk with 7 cm layer of

(*a*) Sand (*b*) Sawdust

(*c*) Dung ash (*d*) All of these

57. Drying of seeds under natural sunlight and storing seeds under well ventilated godowns is

(*a*) Preventive method (*b*) Curative method

(*c*) Both (*a*) & (*b*) (*d*) None of these

58. Most of the stored grain insects die at 55 - 56°C within a period of

(*a*) 3 to 4 minutes (*b*) 4 to 5 minutes

(*c*) 6 to 8 minutes (*d*) 10 to 20 minutes

59. *Anisopteromalus calandrae* is a parasite of

(*a*) *Rhizopertha dominica* (*b*) *Oryzaephilus surinamensis*

(*c*) *Sitophilus oryzae* (*d*) None of these

60. Treatment of grains with 0.2 to 0.5 per cent propionic or acetic acid reduces the

(*a*) Insect population (*b*) Fungus growth

(*c*) Rodent population (*d*) None of these

61. Single dose or acute poisonous chemical to control rat is

(*a*) Zinc phosphide (*b*) Warfarin

(*c*) Difenacoum (*d*) Bromodialone

62. Bromodialone is an anticoagulant use to control rats as

(*a*) Single dose (*b*) Multiple dose

(*c*) Both (*a*) & (*b*) (*d*) None of these

63. The first case of insecticide resistance in India in stored grain pests was that of

(*a*) *Rhyzopertha dominica* (*b*) *Tribolium castaneum*

(*c*) *Caryedon serratus* (*d*) *Oryzaephilus surinamensis*

64. The insecticide used for the management of stored grain pests

(*a*) Chlorpyriphos methyl (Reldan) (*b*) Cyfluthrin (Tempo)

(*c*) Pirimiphos methyl (Actellic) (*d*) All of these

65. Chemosterilant to rodents is

(a) Tepa
(b) Apholate
(c) Glyzophrol
(d) All of these

66. Anticoagulants are antagonistic to

(a) Vitamin A
(b) Vitamin E
(c) Vitamin D
(d) Vitamin K

67. Fumigants use for rodents control are

(a) Cyanogas
(b) Phosphene
(c) Both (a) & (b)
(d) None of these

68. The first report of pesticide resistance from a storage insect pest in India come in 1971 is

(a) *Tribolim castaneum* to DDT and Malathion
(b) *Sitophilus oryzae* and *Rhizopertha dominica* to Lindane, Malathion, Phosphamidon
(c) *Oryzaephilus surinamensis* to Malathion, Lindane
(d) *Trogoderma granarium* to Phosphine

69. Leather beetle, *Dermastes maculates* is originating from

(a) Indonesia
(b) India
(c) Laos
(d) Nepal

70. "Pusa Bin" is fit for safe storage because it is

(a) Moisture proof
(b) Air tight
(c) Resists Thermal Conductivity
(d) All of these

71. The wall of "Pusa Bin" made up of which material that is bad conductors of heat

(a) Burnt Bricks
(b) Unburnt bricks
(c) Mud wall
(d) Concrete

72. Which of the following is an internal feeder?

(a) Rice moth
(b) Khapra beetle
(c) Rice weevil
(d) Red flour beetle

73. Which of the following is an external feeder?

(a) Rice weevil
(b) Lesser grain borer
(c) Rice moth
(d) Angoumois grain moth

74. Webbing of grains is characteristic symptom of

(a) Rice weevil
(b) Lesser grain borer
(c) Rice moth
(d) Pulse beetle

75. Which of the following refers to Bromodiolone poison?

(*a*) Slow acting poison (*b*) Anticoagulant

(*c*) Multi dose poison (*d*) All of these

76. Chlorpyriphos methyl (Reldan) is effective against stored product insects except

(*a*) *Rhyzopertha dominica* (*b*) *Tribolium castaneum*

(*c*) *Caryedon serratus* (*d*) *Oryzaephilus surinamensis*

77. The stored product insect which is more sensitive/susceptible to spinosad is

(*a*) *Rhyzopertha dominica* (*b*) *Tribolium castaneum*

(*c*) *Caryedon serratus* (*d*) *Oryzaephilus surinamensis*

78. Which of the following is a major pest of groundnut?

(*a*) *Sitophilus* (*b*) *Tribolium*

(*c*) *Caryedon* (*d*) *Callosobruchus*

79. For the long term safe storage the moisture content of grains should be

(*a*) 10 -12 per cent (*b*) 9 - 10 per cent

(*c*) 7 - 8 per cent (*d*) 12 - 13 per cent

80. Hidden infestation can be detected by

(*a*) Kjeldal method (*b*) Lowry method

(*c*) Staining method (*d*) Bi-urate method

81. Which of the following is a fumigant used for the management of stored grain insects?

(*a*) Malathion (*b*) Barium Carbonate

(*c*) Methyl bromide (*d*) Zinc phosphide

82. Angoumois grain moth belongs to the family

(*a*) Noctuidae (*b*) Pyralidae

(*c*) Gelechiidae (*d*) Bruchidae

83. Which of the following is a rodenticide of botanical origin?

(*a*) Neem (*b*) Karanja

(*c*) Red squill (*d*) Datura

84. Which of the following is not a above ground storage structure?

(*a*) Gola (*b*) Puri

(*c*) Khatti (*d*) Kothis

85. Disadvantages of Zinc Phosphide baiting is

(a) Necessity of prebaiting

(b) Induce bait shyness

(c) Toxic to non-target species and chances of secondary poisoning are more

(d) All of these

86. Khapra beetle attacks the grain in bulk storage upto a depth of

(a) Inner core (b) Ground level

(c) Peripheral layer of 30 - 40 cms (d) Not attacks

87. Khapra beetle is highly susceptible, particularly in its egg, pupal and adult stages to reduction in

(a) Temperature (b) Moisture

(c) Oxygen tension (d) Light

88. A popular stored grain insect of Indian condition its habit of congregation in cracks and crevices of bricks, masonry and wood storage structures give in the name

(a) Angoumois grain moth (b) Copra beetle

(c) Khapra beetle (d) Carpet beetle

89. Which of the following is a major stored grain pest of groundnut?

(a) *Tribolium* sp. (b) *Caryedon serratus*

(c) *Callosobruchus* spp. (d) *Sitophilus* spp.

90. For the long term safe storage of seeds the moisture content of the seeds should be

(a) 7 to 8 per cent (b) 10 per cent

(c) 11 per cent (d) 12 per cent

91. When aluminium phosphide tablets reacts with moisture then the gas liberated is?

(a) Alethrin (b) Sulphur Di-oxide

(c) Phosphin (d) Oxygen

92. Which of the following pest belongs to the family Dermestidae?

(a) Red flour beetle (b) Flat grain beetle

(c) Khapra beetle (d) Drug store beetle

93. Infestation of most of the insect pests can be avoided if the moisture content below 9 per cent except

(a) Rice moth (b) Rice weevil

(c) Khapra beetle (d) Pulse beetle

94. Circular bore hole in rice is due to

(*a*) Rice moth (*b*) Rice weevil

(*c*) Khapra beetle (*d*) Pulse beetle

95. Which of the following statement is true with respect to anti coagulants?

(*a*) They are safer than acute rodenticides because they are less toxic to the non-target species

(*b*) Quick knock down effect

(*c*) Cost of operation is cheap

(*d*) All of these

96. Which one is not a gelechid?

(*a*) Potato tuber moth (*b*) Angoumois grain moth

(*c*) Rice moth (*d*) Almond moth

97. *Callosobruchus analis* is commonly known as

(*a*) Rice dhora (*b*) Gram dhora

(*c*) Mung dhora (*d*) Puls dhora

98. Which one of the following is a improve storage structure?

(*a*) Godown (*b*) Gola

(*c*) Pusa Bin (*d*) Puri

99. For Vacuum Process Storage (VPS) system__________containers is used.

(*a*) Earthen (*b*) Metal

(*c*) PVC (*d*) None of these

100. Major stored grain pests are belongs to which of the following groups?

(*a*) Diptera and Hemiptera (*b*) Orthoptera and Thysanura

(*c*) Lepidoptera and Coleoptera (*d*) None of these

101. Which of the following pest is not a pest of stored cereals?

(*a*) *Callosobruchus chinensis* (*b*) *Corcyra cephalonica*

(*c*) *Sitotroga cerealella* (*d*) *Sitophilus oryzae*

102. Webbing of rice grains is characteristic symptom of

(*a*) Pulse beetle (*b*) Khapra beetle

(*c*) Rice moth (*d*) Rice weevil

103. In which of the following stored grain pests, infestation starts from the field itself and carried to the godowns

(*a*) *Sitophilus oryzae* (*b*) *Sitotroga cerealella*

(*c*) *Callosobruchus* spp. (*d*) All of these

104. Which of the following pest belongs to the family Dermestidae?

(a) *Sitophilus oryzae* (b) *Trogoderma granarium*

(c) *Tribolium castaneum* (d) None of these

105. Which of the following is a rodenticide of botanical origin?

(a) Red Squill (b) Strychnine

(c) Both (a) & (b) (d) None of these

106. The scientific name of red rust floor beetle is

(a) *Tribolium castaneum* (b) *Trogoderma granarium*

(c) *Rhizopertha domonica* (d) *Sitophilus* sp.

107. Which of the following is an internal feeder?

(a) Pulse beetle (b) Khapra beetle

(c) Rice moth (d) Meal moth

108. *Callosobruchus chinensis* is a primary pest of

(a) Wheat (b) Jowar

(c) Rice (d) Gram

109. *Sitophilus oryzae* belongs to the family

(a) Bruchidae (b) Bostrychidae

(c) Curculionidae (d) Gelechidae

110. *Sitophilus oryzae* lay eggs on

(a) Cracks and crevices (b) Surface of the grain

(c) Cavity makes on grain by adult (d) None of these

111. Pupae is found inside the grains in case of

(a) Rice moth (b) Khapra beetle

(c) Rice weevil (d) None of these

112. Period of activity of *Sitophilus oryzae* is

(a) January to March (b) April to October

(c) March to December (d) None of these

113. Average damage of food grains during storage in India

(a) 14 per cent (b) 12 per cent

(c) 10 per cent (d) 8 per cent

114. *Corcyra cephalonica* is commonly known as

(a) Silk moth (b) Flour beetle

(c) Grain borer (d) Rice moth

115. Which of the following is an underground storage structure?

(*a*) Kothas or Kothis (*b*) Khatti

(*c*) Puri (*d*) Gola

116. Which of the following is an external feeder?

(*a*) Red flour beetle (*b*) Khapra beetle

(*c*) Rice moth (*d*) All of these

117. In which one of the following stored grain pest, only larvae causes damage?

(*a*) Red flour beetle (*b*) Khapra beetle

(*c*) Rice moth (*d*) All of these

118. ________percent of Relative Humidity is ideal for cereal grain storage.

(*a*) 8 - 10 per cent (*b*) 5 - 6 per cent

(*c*) 7 - 9 per cent (*d*) 10 - 15 per cent

119. Which of the following is a secondary storage pest?

(*a*) Flat grain beetle (*b*) Saw toothed grain beetle

(*c*) Long headed flour beetle (*d*) All of these

120. Food grain may be spray with

(*a*) Malathion (*b*) Dichlorovos

(*c*) BHC (*d*) Phorate

121. Which of the following is not a storage pest?

(*a*) Nematodes (*b*) Rodents

(*c*) Insect (*d*) Fungi

122. Bad smell of flour is due to

(*a*) Cockroach (*b*) Khapra beetle

(*c*) Pulse beetle (*d*) Lasser grain borer

123. Which store grain pest require highest temperature?

(*a*) Rice weevil (*b*) Rice moth

(*c*) Khapra beetle (*d*) Pulses beetle

124. Effective control of stored grain pest is

(*a*) Insecticide spray (*b*) Fumigation

(*c*) Cultural practices (*d*) Biological control

125. The best method of control stored grain pests is

(*a*) Fumigation (*b*) Systemic insecticide

(*c*) Proper drying and storage (*d*) Biological agents

126. Value depreciation of a product owing to the present of insects or other foreign matter is called

(*a*) Duty (*b*) Drop

(*c*) Dockage (*d*) None of these

127. Impact machines which are used to kill insects in stored products are also called as

(*a*) Atomizer (*b*) Entoleters

(*c*) Duster (*d*) None of these

128. Which of the following is secondary pest of storage?

(*a*) Long headed flour beetle (*b*) Saw toothed grain beetle

(*c*) Flat grain beetle (*d*) All of these

129. Which of the following pest is an internal feeder?

(*a*) Rice weevil (*b*) Lesser grain borer

(*c*) Angoumois grain moth (*d*) All of these

130. The concentration of the gas per unit volume × time of the exposure is called

(*a*) Contact value (*b*) Crucial value

(*c*) Critical value (*d*) None of these

Fill in the Blanks

1. In the situation of heavy grain moisture content the microorganisms enhance their metabolic activity and generate heat resulting grain heating is called as__________.

2. In the situation of low grain moisture content, but congenial for fast insect development leading to rise in temperature up to 42°C is called as__________.

3. __________is the product of the concentration of the gas per unit volume × time of exposure.

4. Aluminium phosphide tablet has no__________action.

5. In curative measure of stored grain pest use of__________are very effective.

6. Sub-tropical humid and tropical humid climate usually favours__________.

7. Coating by__________is very effective in the management of bruchid in stored pulses and beans.

8. The water present at extra or inner cellular level or on the surface, but not within the cells are known as__________.

9. Losses of uniformity of grain size, appearance, colour, texture etc. are__________loss.

10. Two types of grain losses in store, one is qualitative and another is__________.

11. __________is the balance between grain moisture content and relative humidity in air.

12. For determination of hidden infestation by stored product insects ninhydrin method was first used by__________.

Match the Following

1. Match the pest with their characteristics symptom of damage

(a) Webbing of grains	i) Lesser grain borer
(b) Presence of silken tubes	ii) Rice weevil
(c) Irregular holes on the grains	iii) Rice moth
(d) Circular bore hole in rice	iv) Almond moth

2. Match the pest with their characteristics symptom of damage

(a) Grains covered with scales	i) Angoumois grain moth
(b) Damaged grains and webbing	ii) Red flour beetle
(c) Circular holes or exit holes	iii) Pulse beetle
(d) Powdery grains and foul smell	iv) Indian meal moth

3. Match the pest with their scientific name

(a) Almond moth	i) *Plodia interpunctella*
(b) Indian meal moth	ii) *Euphestia cautella*
(c) Angoumois grain moth	iii) *Phthorimaea operculella*
(d) Potato tuber moth	iv) *Sitotroga cerealella*

4. Match the pest with their scientific name

(a) Red rust flour beetle	i) *Tribolium castaneum*
(b) Khapra beetle	ii) *Sitophilus oryzae*
(c) Lesser grain borer	iii) *Trogoderma granarium*
(d) Rice weevil	iv) *Rhizopertha domonica*

5. Match the pest with their scientific name

(a) Pulse beetle	i) *Callosobruchus chinensis*
(b) Rice moth	ii) *Oryzaephilus surinamensis*
(c) Saw toothed grain beetle	iii) *Corcyra cephalonica*
(d) Carpet beetle	iv) *Anthrenus flavipes*

6. Match the following pests with their family

(a)	Pulse beetle	i)	Gelechidae
(b)	Lesser grain borer	ii)	Bruchidae
(c)	Rice weevil	iii)	Bostrychidae
(d)	Rice moth	iv)	Curculionidae

Answers

Multiple Choice Questions

1	(a)	21	(a)	41	(a)	61	(a)
2	(c)	22	(b)	42	(c)	62	(c)
3	(a)	23	(a)	43	(c)	63	(b)
4	(b)	24	(a)	44	(a)	64	(d)
5	(d)	25	(d)	45	(c)	65	(d)
6	(c)	26	(a)	46	(a)	66	(d)
7	(b)	27	(b)	47	(a)	67	(c)
8	(c)	28	(a)	48	(a)	68	(a)
9	(b)	29	(b)	49	(c)	69	(b)
10	(d)	30	(a)	50	(c)	70	(d)
11	(d)	31	(d)	51	(c)	71	(b)
12	(d)	32	(c)	52	(b)	72	(c)
13	(a)	33	(c)	53	(b)	73	(c)
14	(a)	34	(b)	54	(b)	74	(c)
15	(a)	35	(c)	55	(c)	75	(d)
16	(b)	36	(d)	56	(d)	76	(a)
17	(b)	37	(d)	57	(c)	77	(a)
18	(a)	38	(c)	58	(d)	78	(c)
19	(d)	39	(b)	59	(c)	79	(c)
20	(a)	40	(b)	60	(b)	80	(c)

81	(c)	94	(b)	107	(a)	120	(a)
82	(c)	95	(a)	108	(d)	121	(a)
83	(c)	96	(d)	109	(c)	122	(a)
84	(c)	97	(c)	110	(c)	123	(c)
85	(d)	98	(c)	111	(c)	124	(b)
86	(c)	99	(c)	112	(b)	125	(a)
87	(c)	100	(c)	113	(c)	126	(c)
88	(c)	101	(a)	114	(d)	127	(b)
89	(b)	102	(c)	115	(b)	128	(d)
90	(a)	103	(d)	116	(d)	129	(d)
91	(c)	104	(b)	117	(c)	130	(c)
92	(c)	105	(c)	118	(c)		
93	(c)	106	(a)	119	(d)		

Fill in the Blanks

1. Damp grain heating
2. Dry grain heating
3. Critical value
4. Ovicidal
5. Fumigants
6. Field infestation
7. Vegetable oils
8. Free water
9. Qualitative
10. Quantitative
11. Relative humidity equilibrium pattern
12. Dennis and Decker (1976)

Match the Following

1. a) - iii, b) - iv, c) - i, d) - ii
2. a) - i, b) - iv, c) - iii, d) - ii
3. a) - ii, b) - i, c) - iv, d) - iii
4. a) - i, b) - iii, c) - iv, d) - ii
5. a) - i, b) - iii, c) - ii, d) - iv
6. a) - ii, b) - iii, c) - iv, d) - i

Chapter 13

Arthropod Vectors of Plant Diseases

Multiple Choice Questions

1. **Virus transmission by plant hopper in__________manner.**
   (*a*) Circulative and Propagative (*b*) Persistant and Circulative
   (*c*) Semi-persistant (*d*) None of these

2. **Which virus is a type of parasite virus?**
   (*a*) Obligate (*b*) Facultative
   (*c*) Both (*a*) & (*b*) (*d*) None of these

3. **Rice tungro disease is transmitted by**
   (*a*) White flies (*b*) Brown Plant Hopper
   (*c*) Aphids (*d*) None of these

4. **The vector of Bean mosaic and Leaf crinkle of Urad is**
   (*a*) *Aphis craccivora* (*b*) *Bemisia tabaci*
   (*c*) *Pentalonia nigronervosa* (*d*) *Myzus persicae*

5. **Virus of Cacao mottle is transmitted by**
   (*a*) Mealy bug (*b*) Aphid
   (*c*) Hoppers (*c*) Mite

6. **Virus transmission by white fly is persistent and**
   (*a*) Circulative type (*b*) Not propagative type
   (*c*) Mechanical type (*d*) Non mechanical type

7. The vector of Chilli mosaic disease is

(a) *Aphis craccivora* (b) *Bemisia tabaci*

(c) *Pentalonia nigronervosa* (d) *Myzus persicae*

8. Malformation is common in all parts of plants in the form of tumour and galls due to stimulus caused by phytotoxemia at

(a) Tissue level (b) Organ level

(c) Organism level (d) Cellular level

9. The virus is retained through the moult of the insect a feature which is called__________transmission.

(a) Trans-stadial (b) Trans-ovarial

(c) Bimodal (d) Trimodal

10. The circulative type of viruses are not carried in the styletes and they reach to midgut, and the transmission is

(a) Non mechanical (b) Mechanical

(c) Aerial (d) None of these

11. Soybean Mild Mosaic Virus is transmitted by

(a) White fly (b) Aphid

(c) Mealy bug (d) Thrips

12. Virus transmission by mites is

(a) Circulative but not propagative (b) Semi persistent

(c) Non persistent (d) All of these

13. Vector of little leaf of brinjal is

(a) Leaf hopper (b) Plant hopper

(c) White fly (d) Mealy bugs

14. Propagative viruses sometimes transmitted through the eggs of the infected vector to its progeny known as

(a) Trans-stadial transmission (b) Trans-ovarial transmission

(c) Bimodal transmission (d) Trimodal

15. Piercing and sucking type of mouthparts of insects makes them effective as__________of plant viruses.

(a) Natural enemy (b) Vectors

(c) Predators (d) Parasitoids

16. Which of the following is an example of potato virus Y?

(a) Leaf rot (b) Rugose mosaic

(c) Mild mosaic (d) Vein clearing

17. Biological agencies, transmitted the pathogens of plant disease are known as

(a) Bio-control
(b) Vectors
(c) Entomopathogen
(d) Virus

18. The varying lengths of time, the viruses spend in their insect vectors before resumption of activity in the plant hosts, commonly known as_________ period.

(a) Inoculation
(b) Contamination
(c) Incubation or Latent
(d) Introduction

19. The vector is able to acquire the virus from a disease source and transmit it to a healthy plant and the vector losses its infectivity, such types of virus are called_________viruses.

(a) Non circulative
(b) Non persistent
(c) Stylet borne
(d) All of these

20. The virus are carried in the stylets and they doesn't reach midgut is known as, stylet borne viruses and the type of transmission is

(a) Circulative
(b) Persistent
(c) Mechanical
(d) Aerial

21. The vector remains to be viruliferous for rest of its life i.e. the infectivity lasts until its death and no need to feed regularly whenever it transmit virus are known as_________viruses.

(a) Circulative type
(b) Persistent
(c) Non stylet borne
(d) All of these

22. Red gram sterility mosaic virus is transmitted by

(a) Mites
(b) Aphids
(c) Mealy bugs
(d) White fly

23. Some circulative viruses may multiply in the bodies of their respective vectors, so they are called as_________viruses, transmitted by some leaf hoppers.

(a) Propagative
(b) Mechanical
(c) Aerial
(d) Non propagative

24. Cabbage black ring spot is transmitted only by

(a) *Aphis gossipy*
(b) *Lipaphis erisimi*
(c) *Myzus persicae*
(d) *Aphis craccivora*

25. Homopteran insects alone are reported to transmit about_________percent of the known plant viruses.

(a) 60
(b) 70
(c) 80
(d) 90

26. Among homopteran the predominant vectors are

(*a*) Mealy bug (*b*) Aphids

(*c*) Coccids (*d*) Scale insects

27. Chemicals damaging to plant life are known as

(*a*) Antifeedant (*b*) Phytotoxic

(*c*) Pheromone (*d*) Antixenosis

28. The toxic effects to plants induced by salivary component secreted by the arthropod specially insects while feeding referred as

(*a*) Antixenosis (*b*) Antibiosis

(*c*) Phytotoxemia (*d*) Plantomycine

29. Due to effect of phytotoxemia tea mosquito bug *Helopeltis theivora* produces

(*a*) Blue lesion on leaf (*b*) Green lesion on leaf

(*c*) Yellow lesion on leaf (*d*) Red lesion on leaf

30. *Chaetocnema pulicaria*, which is a vector of bacterial wilt of corn is a

(*a*) White fly (*b*) Aphid

(*c*) Butterfly (*d*) Beetle

31. Vector of cotton wilt, *Melanoplus differentialis*, which is a/an

(*a*) Grasshopper (*b*) White fly

(*c*) Aphid (*d*) None of these

32. Tomato leaf curl virus is transmitted by

(*a*) Grasshopper (*b*) White fly

(*c*) Aphids (*d*) Mealy bugs

33. Some vectors (aphid) transmit non-circulative viruses either in non-persistant or semi-persistant manner this method is known as__________transmission.

(*a*) Trans-stadial (*b*) Trans-ovarial

(*c*) Bimodal (*d*) Trimodal

34. The phenomenon where an insect transmitted disease of an animal species enters a human population is known as

(*a*) Trans-stadial (*b*) Trans-ovarial

(*c*) Bimodal (*d*) Zoonosis

35. Virus transmission by mealy bugs is

(*a*) Circulative (*b*) Semi persistent

(*c*) Non persistent (*d*) None of these

36. The vector of Tomato spotted wilt is

(a) Grasshopper (b) Thrips
(c) Aphid (d) Mealy bugs

37. Peanut Bud Necrosis Disease (PBND) virus is transmitted by

(a) Thrips (b) Aphids
(c) Whitefly (d) All of these

38. Soybean Yellow Mosaic Virus is transmitted by

(a) Whitefly (b) Aphids
(c) Butterfly (d) Beetle

39. Persistent and circulative manner of virus transmission is found

(a) Mealy bug (b) Aphid
(c) Hoppers (d) Thrips

40. Vector of Rice Tungro disease is

(a) Green Leaf Hopper (b) Brown Plant Hopper
(c) White backed plant hopper (d) All of these

41. Papaya leaf curl virus is transmitted by

(a) Aphid (b) Thrips
(c) White fly (d) Mealy bug

42. Which one of the following is a major vector of plant disease?

(a) Mustard saw fly (b) Fruit fly
(c) Cotton white fly (d) Sugarcane top borer

43. Which one of the following function (s) of mouth parts of insects makes them effective vectors of plant viruses?

(a) Biting and Chewing (b) Piercing and Sucking
(c) Rasping (d) Sponging

44. Rice Ragged Stunt is transmitted by

(a) *Nilaparvata lugens* (b) *Nephotettix virescens*
(c) *Sogatodes oryzicola* (d) All of these

45. The vector of Banana Bunchy top disease is

(a) *Aphis craccivora* (b) *Bemisia tabaci*
(c) *Pentalonia nigronervosa* (d) *Myzus persicae*

46. The vector of Yellow vein mosaic disease of Bhindi is

(a) *Aphis craccivora* (b) *Bemisia tabaci*
(c) *Pentalonia nigronervosa* (d) *Myzus persicae*

47. Peanut stripe virus is transmitted by

(*a*) Thrips (*b*) Aphids

(*c*) Whitefly (*d*) None of these

48. Wheat streak mosaic virus is transmitted by

(*a*) Mites (*b*) Aphids

(*c*) Cut worms (*d*) Termites

49. Yellow mosaic viral disease of moong is transmitted by

(*a*) *Nephotettix* spp. (*b*) *Bemesia tabaci*

(*c*) *Aphis craccivora* (*d*) *Amrasca bigutula*

50. Green leaf hopper of rice is the primary vector of

(*a*) Tungro (*b*) Rice dwarf

(*c*) Rice yellow dwarf (*d*) All of these

51. The feeding time necessary to acquire a sufficient titer of virus by an insect vector is called

(*a*) Incubation period (*b*) Latent period

(*c*) Acquisition period (*d*) Dormant period

Fill in the Blanks

1. Biological agencies, which transmits pathogens of disease called__________.

2. Subsequent to acquisition of virus the vector became infective only after a certain period is called as__________.

3. Cabbage black ring spot is transmitted by__________.

4. Sandal spike disease is caused by phytoplasma and transmitted by the insect vector__________.

5. Stewart's disease of corn is a bacterial disease, transmitted by flea beetle__________.

Match the Following

1. Match the following Mycoplasmal diseases of plant and their vectors

(*a*) Rice Yellow Dwarf	i) *Cestius phycitis* (Hopper)
(*b*) Little leaf of brinjal	ii) *Stephanitis typica*
(*c*) Coconut Root Wilt	iii) *Nephotettix virescens* (GLH)
(*d*) Sesamum phyllody	iv) *Orosius albicinctus* (Leafhopper)

2. **Match the following diseases of plant and their vectors**

(a) Rice tungro	i) Eriophyid mite
(b) Cotton leaf curl	ii) Green leafhopper
(c) Wheat streak mosaic	iii) Whitefly
(d) Sesamum phyllody	iv) Leaf hopper

3. **Match the following diseases of plant and their vectors**

(a) Katte disease of cardamom	i) Thrips
(b) Chilli leaf curl	ii) Whitefly
(c) Peanut bud necrosis virus	iii) Aphid
(d) Banana bunchy top	iv) Banana aphid

4. **Match the following diseases of plant and their vectors**

(a) Peanut bud necrosis virus	i) Psyllid
(b) Citrus greening	ii) Aphid
(c) Citrus tristeza virus	iii) Thrips
(d) Okra yellow vein mosaic	iv) Whitefly

5. **Match the type of virus transmission with their insect vectors**

(a) Non-persistent	i) Aphids
(b) Semi-persistent	ii) Leafhoppers
(c) Non propagative	iii) Whitefly
(d) Propagative	iv) Thrips

6. **Match the insect vectors with their type of virus transmission**

(a) White fly	i) Begomovirus
(b) Aphids	ii) Tospovirus
(c) Thrips	iii) Potyviridae
(d) Mealy bugs	iv) Badnavirus

7. **Match the following terms with their explanations**

(a) Vector	i) Biological agency transmit disease pathogen
(b) Phytotoxemia	ii) Physiological changes of plant due to chemical effect
(c) Zoonosis	iii) Ultramicroscopic in size, live and multiply only in living cells
(d) Virus	iv) Insect transmitted disease of an animal species enters to a man

Answers

Multiple Choice Questions

1	(a)	14	(b)	27	(b)	40	(a)
2	(a)	15	(b)	28	(c)	41	(c)
3	(d)	16	(b)	29	(d)	42	(c)
4	(a)	17	(b)	30	(d)	43	(b)
5	(a)	18	(c)	31	(a)	44	(a)
6	(a)	19	(d)	32	(b)	45	(c)
7	(b)	20	(c)	33	(c)	46	(b)
8	(d)	21	(d)	34	(d)	47	(b)
9	(a)	22	(a)	35	(b)	48	(a)
10	(a)	23	(a)	36	(b)	49	(c)
11	(b)	24	(c)	37	(a)	50	(d)
12	(a)	25	(d)	38	(a)	51	(c)
13	(a)	26	(b)	39	(d)		

Fill in the Blanks

1. Vectors
2. Incubation period or Latent period
3. *Myzus persicae*
4. *Jassus indicus* (Hopper)
5. *Chaetocnema* spp.

Match the Following

1. a) - iii, b) - i, c) - ii, d) - iv
2. a) - ii, b) - iii, c) - ii, d) - iv
3. a) - iii, b) - ii, c) - i, d) - iv
4. a) - iii, b) - i, c) - ii, d) - iv
5. a) - i, b) - ii, c) - iii, d) - iv
6. a) - i, b) - iii, c) - ii, d) - iv
7. a) - i, b) - ii, c) - iv, d) - iii

Chapter 14

Honey Bees and Beekeeping

Multiple Choice Questions

1. **Giant Rock bee (*Apis dorsata*) belongs to the family**
   (*a*) Apinae (*b*) Apidae
   (*c*) Vespidae (*d*) None of these

2. **Dammer bee or mosquito bee, *Trigona* (*Melipona*) *iridipennis* belongs to the family**
   (*a*) Apinae (*b*) Apidae
   (*c*) Vespidae (*d*) Meliponidae

3. **The cap of the workers cell is**
   (*a*) Concave (*b*) Convex
   (*c*) Flat (*d*) Irregular

4. **Vestigial stings are found in**
   (*a*) Giant Rock bee (*b*) Little bee
   (*c*) Indian bee and Italian bee (*d*) Dammer bee

5. **Round dance indicate that source of food is**
   (*a*) Less than 50 m (*b*) More than 50 m
   (*c*) Less than 100 m (*d*) More than 100 m

6. **The ascending order of the size of the cell is**
   (*a*) Worker $<$ Drone $<$ Queen (*b*) Worker $>$ Drone $>$ Queen
   (*c*) Drone $<$ Worker $<$ Queen (*d*) Queen $>$ Drone $>$ Worker

7. Which of the following is not an Indian bee

(*a*) *Apis dorsata* (*b*) *Apis indica*

(*c*) *Apis florae* (*d*) *Apis mellifera*

8. The mixture of wax and earth or resin is known as

(*a*) Nectar (*b*) Cerumen

(*c*) Bee Bread (*d*) Propolis

9. Dammer bee or mosquito bee, *Trigona* (*Melipona*) *iridipennis* made up their comb with

(*a*) Earth (*b*) Proplis

(*c*) Cerumen (*d*) None of these

10. Which of the following chemicals are found in royal jelly?

(*a*) Lipoproteins (*b*) Natural glycerides

(*c*) Vitamins (*d*) All of these

11. Which type of antenna honey bee have?

(*a*) Aristate (*b*) Geniculate

(*c*) Pectinate (*d*) None of these

12. Honey bee have which type of mouth parts

(*a*) Chewing and Biting (*b*) Chewing and lapping

(*c*) Siphoning (*d*) Sucking

13. Propodeum is the fusion of 1st abdominal segment with

(*a*) 2nd abdominal segment (*b*) Prothorax

(*c*) Mesothorax (*d*) Metathorax

14. Stingless honey bee is

(*a*) Queen Daughter (*b*) Worker

(*c*) Drones (*d*) Queen Mother

15. Royal jelly is produced by

(*a*) Workers (*b*) Queen

(*c*) Drones (*d*) Sterile male

16. A bee colony is considered week or strong according to the number of

(*a*) Workers (*b*) Queen

(*c*) Drones (*d*) All of these

17. Honey bee queen lay about 15 lakh eggs in her life time and per day lay

(*a*) 100 - 400 eggs (*b*) 400 - 800

(*c*) 800 - 1200 (*d*) 1200 - 1600

18. Drones are produced from

(*a*) Fertilized eggs (*b*) Unfertilized eggs

(*c*) ***Both (a) & (b)*** (*d*) None of these

19. Who are not permanent members of bee colony?

(*a*) Queen Daughter (*b*) Worker

(*c*) Drones (*d*) Queen Mother

20. Workers are produced from

(*a*) Fertilized eggs (*b*) Unfertilized eggs

(*c*) ***Both (a) & (b)*** (*d*) None of these

21. The largest bee or rock bee which makes largest hive

(*a*) *Apis dorsata* (*b*) *Apis indica*

(*c*) *Apis florae* (*d*) *Apis mellifera*

22. The most ferocious bee is

(*a*) *Apis dorsata* (*b*) *Apis indica*

(*c*) *Apis florae* (*d*) *Apis mellifera*

23. The cap of the drone cell is__________with a central hole.

(*a*) Concave (*b*) Convex

(*c*) Flat (*d*) Irregular

24. Queen cells are situated in the brood comb at

(*a*) Lower border (*b*) Middle row

(*c*) Upper border (*d*) None of these

25. Queen daughter take nuptial flight at after emergence of

(*a*) 2 - 3 days (*b*) 4 - 5 days

(*c*) 7 - 8 days (*d*) 9 -10 days

26. The Indian honey bee is technically named as

(*a*) *Apis dorsata* Fabricious (*b*) *Apis indica* Fabricious

(*c*) *Apis florae* Fabricious (*d*) *Apis mellifera* Linn.

27. Wag tail dance indicate that source of food is

(*a*) Very near (*b*) Near

(*c*) Longer distance (*d*) None of these

28. Pollen baskets of the worker bee located in

(*a*) Hind tibiae (*b*) Fore tibia

(*c*) Middle tibia (*d*) None of these

29. In honey bees four pairs of wax glands are located on

(*a*) 3 - 6 abdominal sterna (*b*) 4 - 7 abdominal terga
(*c*) 3 - 6 abdominal pluron (*d*) None of these

30. The fully ripened honey contains the highest amount of

(*a*) Water (*b*) Glucose
(*c*) Fructose (*d*) Sucrose

31. Honey bees may drink the total honey of the colony before

(*a*) Migration (*b*) Absconding
(*c*) Honey collection (*d*) Nuptial flight

32. The number of queens present in a bee colony

(*a*) One (*b*) Two
(*c*) Three (*d*) Same numbers as drones

33. Who discovered the bee language (bee dance) in early 1920s and was awarded the Noble Prize in 1973?

(*a*) Charls Elton (*b*) Prof. Karl von Frisch
(*c*) A. Bassi (*d*) AD Imms

34. Rearing of the bees in artificial hives is known as

(*a*) Bee culture (*b*) Beekeeping
(*c*) Honey rearing (*d*) Honey culture

35. Modern method of Beekeeping is also known as

(*a*) Cage culture (*b*) Frame Hive method
(*c*) Common method (*d*) Comb method

36. In India first small hive was designed by father Newton in 1910 for

(*a*) Apis *dorsata* (*b*) *Apis cerana indica*
(*c*) *Apis melifera* (*d*) *Trigona* (*Melipona*) *iridipennis*

37. The plants that yield nectar and pollen are collectively called

(*a*) Bee flora (*b*) Bee pasturage
(*c*) Bee forage (*d*) All of these

38. The days of without honey flow are called

(*a*) Lack period (*b*) Dead period
(*c*) Dearth period (*d*) Dry period

39. Nosema disease of honey bee caused by *Nosema apis* is a

(*a*) Bacteria (*b*) Protozoa
(*c*) Fungus (*d*) Virus

40. In honey bee fertilized eggs produces

(*a*) Females (*b*) Males

(*c*) Both Males and Females (*d*) None of these

41. A new drug introduce to control Nosema disease

(*a*) Gentamycine (*b*) Fumagillin

(*c*) Ampicillin (*d*) None of these

42. Amoebic disease of honey bee caused by *Malpighamoeba mellificae*

(*a*) Bacteria (*b*) Protozoa

(*c*) Fungus (*d*) Virus

43. The behavior of *Apis mellifera* is similar to

(*a*) *Apis dorsata* (*b*) *Apis indica*

(*c*) *Apis florae* (*d*) None of these

44. Europian foulbrood is caused by a bacteria

(*a*) *Bacillus larvae* (*b*) *Melisscoccus pluton*

(*c*) Both (*a*) & (*b*) (*d*) None of these

45. Chalkbrood disease of honey bee is caused by *Ascophaera apis* is a

(*a*) Bacteria (*b*) Protozoa

(*c*) Fungus (*d*) Virus

46. Sacbrood disease of honey bee larvae is caused by

(*a*) Bacteria (*b*) Protozoa

(*c*) Black Fungus (*d*) Small RNA virus

47. Acarine disease or Isle of wight disease of adult honey bee caused by *Acarapis woodi* is a

(*a*) Bacteria (*b*) Protozoa

(*c*) Fungus (*d*) Mite

48. The origin of European bee (*Apis mellifera*) is

(*a*) Italy (*b*) Spain

(*c*) Germany (*d*) France

49. Scientific name of Greater wax moth is

(*a*) *Achroia grisella* (*b*) *Galleria mellonella*

(*c*) *Vespa magnifica* (*d*) *Vespa orientalis*

50. Scientific name of lesser wax moth is

(*a*) *Achroia grisella* (*b*) *Galleria mellonella*

(*c*) *Vespa magnifica* (*d*) *Vespa orientalis*

51. The larvae of the wax moths attacks

(a) The larva of Honey bee
(b) The combs and makes silken tunnels
(c) Drinks honey
(d) None of these

52. Bird enemy of honey bee is

(a) *Componotus compressus* (b) *Dorylus labiatus*
(c) Monomorium indicum (d) *Merops orientalis*

53. The smallest bee among the following is

(a) *Apis dorsata* (b) *Apis indica*
(c) *Apis florae* (d) *Apis mellifera*

54. The stingless bees belong to the family

(a) Bombidaae (b) Meliponidae
(c) Apidae (d) Aphididae

55. National Beekeeping Development Board was established in

(a) 1993 (b) 1994
(c) 1995 (d) 1996

56. National Beekeeping Development Board now renamed as

(a) National Bee Board (b) Bee Board of India
(c) Indian Bee Board (d) Bee Monitoring Board

57. Indian Bee Journal is published by

(a) Beekeepers Association of India
(b) All India Beekeepers Association
(c) Indian Beekeepers Association
(d) AICRP on Honey Bee

58. The bee which makes single vertical comb

(a) *Apis dorsata* (b) *Apis florae*
(c) *Apis indica* (d) Both (a) & (b)

59. Wax gland are found on

(a) Ventral side of abdomen (b) Dorsal side of abdomen
(c) Whole Body (d) None of these

60. For pollen collection which legs are modified

(a) Hind legs (b) Fore legs
(c) Middle legs (d) Not modified

61. The bee which makes series of parallel combs

(*a*) *Apis dorsata* (*b*) *Apis indica*

(*c*) *Apis mellifera* (*d*) Both (*b*) & (*c*)

62. Central Bee Research and Training Institute (CBRTI) is situated at

(*a*) Ranchi (*b*) Ludhiana

(*c*) Pune (*d*) New Delhi

63. Large size single comb is made by

(*a*) *Apis cerana indica* (*b*) *Apis melifera*

(*c*) *Trigona* (*Melipona*) *iridipennis* (*d*) *Apis dorsata*

64. *Acarapis woodi* live, feed and breed in the

(*a*) Tracheae of the adult bee (*b*) Alimentary canal

(*c*) Malpighian tubules (*d*) None of these

65. The infected larvae by American foulbrood grow to almost full size but die of

(*a*) Diarrhoea (*b*) Septicemia

(*c*) Dysentery (*d*) Swelling

66. Bee dance depends upon distance of food source and direction of

(*a*) Sun light (*b*) Colour of flower

(*c*) Water source (*d*) Wind

67. *Nosema apis* attacks

(*a*) The lining of the stomach (*b*) Nervous system

(*c*) Cuticle (*d*) Eyes

68. Stinging organ of honey bee is the modified

(*a*) Genetelia (*b*) Anal Style

(*c*) Anal Cerci (*d*) None of these

69. Co-operation among individuals, care of youngs, division of labour, sterile individuals assists fecund individuals, an overlap of generations so that off springs are able to assists their parents are called as

(*a*) Eusocial insects (*b*) Social insects

(*c*) Semi Social insects (*d*) Communal insects

70. Mandibles of worker honey bee are flattened and spoon shaped because of

(*a*) Necter collection

(*b*) Molding the wax for comb building

(*c*) For biting

(*d*) None of these

71. The main honey producing bee family

(*a*) Bombidaae (*b*) Meliponidae
(*c*) Apidae (*d*) Aphididae

72. Which of the following colour honey bee can distinguish?

(*a*) Black (*b*) Red
(*c*) Grey (*d*) Yellow

73. The largest honey bee is

(*a*) *Apis dorsata* (*b*) *Apis cerana indica*
(*c*) *Apis melifera* (*d*) *Trigona* (*Melipona*) *iridipennis*

74. Annual honey yield of a *Apis dorsata* colony is

(*a*) 35 - 40 kg (*b*) 10 - 15 kg
(*c*) 16 - 25 kg (*d*) 25 - 30 kg

75. The invertase contained in the saliva acts upon sucrose of the nectar and converts it to

(*a*) Maltose or Dextrose (*b*) Dextrose or Galctose
(*c*) Levulose or Dextrose (*d*) Maltose or Fructose

76. The bees that go out first to find out new sources of pollen, nectar, propolis are called

(*a*) Mentor bee (*b*) Scout bees
(*c*) Collector bee (*d*) Escort bee

77. Cell arrangement of bee hive from upper border to lower border are as follows

(*a*) Honey chamber - Worker brood cells - Pollen cells - Drone and Queen cells
(*b*) Pollen cells - Honey chamber - Worker brood cells - Drone and Queen cells
(*c*) Honey chamber - Pollen cells - Worker brood cells - Drone and Queen cells
(*d*) Honey chamber - Pollen cells - Drone and Queen cells - Worker brood cells

78. Swarming in honeybee is due to

(*a*) Overcrowding (*b*) Scarcity of food
(*c*) Unfavorable environment (*d*) Disturbance

79. The maximum honey producing state is

(*a*) Uttar Pradesh (*b*) Bihar
(*c*) Punjab (*d*) Haryana

80. Which of the following books are authored by Prof. Karl von Frisch?

(*a*) Dance Language and Orientation of bees (*b*) Bees : Their Vision
(*c*) Chemical Sense and Language (*d*) All of these

81. Haploid bee of bee colony is

(*a*) Worker (*b*) Queen

(*c*) Drone (*d*) All of these

82. A pure sample of honey should granulate homogenously and should weight not less than__________per litre.

(*a*) 1.2 kg (*b*) 2.2 kg

(*c*) 1.4 kg (*d*) 1.5 kg

83. The species of honeybees that are fully domesticated

(*a*) Indian bee (*b*) Italian bee

(*c*) Both (*a*) & (*b*) (*d*) None of these

84. The average hive temperature of honey bee is

(*a*) 33 - 36 °C (*b*) 28 - 30 °C

(*c*) 38 - 40 °C (*d*) None of these

85. The grooming of a new queen in the honey bee colony depends upon the secretion known as

(*a*) Royal jelly (*b*) Queen substance

(*c*) Sucrose (*d*) Honey

86. Leaving of bee with queen is known as

(*a*) Swarming (*b*) Absconding

(*c*) Queen excluder (*d*) None of these

87. In honey bees, when the larvae are feed on regular diet of pollen and honey which caste will be produced

(*a*) Worker (*b*) Drone

(*c*) Queen (*d*) All of these

88. The gland which produces pheromone in queen honey bee

(*a*) Mandibular gland (*b*) Maxillary gland

(*c*) Labial gland (*d*) Pharyngeal gland

89. Queen substance secretes by queen bees and this checks the development of__________in workers.

(*a*) Ovaries (*b*) Testis

(*c*) Malpighian tubule (*d*) Gizzard

90. The replacement of the old queen by the new daughter queen is known as

(*a*) Swarming (*b*) Absconding

(*c*) Queen excluder (*d*) Supersedure

91. Up to which age (days), Royal jelly is produced by honey bee worker

(*a*) 5 (*b*) 10

(*c*) 15 (*d*) 20

92. Royal Jelly is reached with

(*a*) Carbohydrate (*b*) Protein

(*c*) Fat (*d*) Ash

93. In India (from a single comb) more honey is obtained from which bee?

(*a*) Rock bee (*b*) Indian bee

(*c*) Italian bee (*d*) Little bee

94. The disease in honeybees which attacks the lining of the stomach causing dysentery

(*a*) Nosema disease (*b*) Amoebic disease

(*c*) Acarine disease (*d*) Brood disease

95. Where the nectar and pollen is converted into honey in the workers?

(*a*) Accessory gland (*b*) Alimentary canal

(*c*) Malpighian tubule (*d*) Wax gland

96. Which one of the following bee hives is more popular in India?

(*a*) Dadant hive (*b*) Newton hive

(*c*) Langstroth hive (*d*) Thompson hive

97. Among total solids percentage of total sugar in matured honey is

(*a*) 62 per cent (*b*) 69 per cent

(*c*) 73 per cent (*d*) 95 per cent

98. One of the major insect pests of honey bees are

(*a*) *Eublema* sp. (*b*) *Vespa cincta*

(*c*) *Formicaria rufa* (*d*) *Galleria mellonella*

99. The worker bee is

(*a*) Sterile female (*b*) Sterile male

(*c*) Fertile male (*d*) Fertile female

100. Queen substance is secreted by

(*a*) Pharyngeal gland (*b*) Mandibular gland

(*c*) Silk gland (*d*) Wax gland

101. The amount of sucrose in fully ripened honey is

(*a*) > 5 per cent (*b*) > 10 per cent

(*c*) < 2 per cent (*d*) > 2 percent

102. Bee venom is used for__________.

(*a*) Curing renal problems (*b*) Curing heart problems

(*c*) Curing body pains (*d*) Curing pains at joints

103. The immature stage of honey bee queen feed on

(*a*) Honey (*b*) Nectar

(*c*) Pollen (*d*) Royal jelly

104. Flowering plant recommended for apiary is

(*a*) Rose (*b*) Tulip

(*c*) Jasmine (*d*) Chrysanthemum

105. Royal jelly is the secretion of

(*a*) Labial glands (*b*) Mandibular glands

(*c*) Neurosecretary cells (*d*) Hypopharyngeal glands

106. Bee milk or brood food is also known as

(*a*) Propolis (*b*) Royal jelly

(*c*) Pollen (*d*) All of these

107. Honey stomach is also called as

(*a*) Filter chamber (*b*) Proventriculus

(*c*) Caecae (*d*) Colon

108. Which of the following are pest of honey bee

(*a*) *Vespa cincta* (*b*) *Palerus orientalis*

(*c*) *Galleria melonella* (*d*) All of these

109. The species of honeybees that not fully domesticated

(*a*) Rock bee (*b*) Little bee

(*c*) Both (*a*) & (*b*) (*d*) None of these

110. Moveable frame hive was first introduced by L. L. Langstroth in

(*a*) 1751 (*b*) 1800

(*c*) 1851 (*d*) 1881

111. The part of hive which allows workers to pass through it easily but retains the queen in brood chamber is known as

(*a*) Swarming (*b*) Absconding

(*c*) Queen excluder (*d*) Supersedure

112. Bee venom is secreted from

(*a*) Wax gland (*b*) Nectar gland

(*c*) Acid gland (*d*) None of these

113. In honey bees, when the larvae are fed on regular diet of pollen and honey, which cast is produced

(*a*) Queen (*b*) Drones

(*c*) Workers (*d*) None of these

114. Sting in worker honey bee is a modified part of

(*a*) Ovipositor (*b*) Anal cerci

(*c*) Mouth parts (*d*) Antenna

115. Where the nectar is converted into honey?

(*a*) In the alimentary canal of queen (*b*) In the alimentary canal of worker

(*c*) In royal chamber (*d*) In special cells of the hive

116. Acute bee paralysis is caused by

(*a*) Bacteria (*b*) Virus

(*c*) Fungus (*d*) Protozoa

117. The disease caused in honeybees by a bacterium, *Bacillus larvae*, is known as

(*a*) American foul brood (*b*) European foul brood

(*c*) Chalk brood (*d*) Sac brood

118. Father of modern beekeeping in India

(*a*) A S Atwal (*b*) A T Atwal

(*c*) A L Atwal (*d*) A P Atwal

119. Which is not secreted by the honey bee?

(*a*) Bees wax (*b*) Royal jelly

(*c*) Propolis (*d*) Queen substance

Fill in the Blanks

1. __________is an antioxidant found only in honey.

2. __________is the place where honey bees are kept.

3. __________is the liquid substances of the flowers and provides carbohydrate source for energy.

4. Approximate larval duration of queen bees are__________days.

5. Worker honey bees cannot fly below__________temperature

6. Propolis is also known as__________which is the sticky substance of bee hive.

7. The mixture of honey and pollen is called as__________.

8. Sac brood virus of honey bee infects the__________portion of body.

9. American foul brood bacteria infect the__________stage of honey bees.

10. In honey bees mating takes place in air during__________.

Match the Following

1. Match the common name and scientific names of honey bees

(a) *Apis melifera*	i) Rock bee
(b) *Apis cerana indica*	ii) Dammer bee
(c) *Melipona irridipennis*	iii) Indian bee
(d) *Apis dorsata*	iv) Italian bee

2. Match the common name and scientific names of honey bees

(a) *Apis melifera*	i) 250 g per colony
(b) *Apis florae*	ii) 20 - 25 kg per colony
(c) *Apis cerana indica*	iii) 3 - 4 kg per colony
(d) *Apis dorsata*	iv) 20 - 40 kg per colony

3. Match the causal organism and disease of honey bee

(a) *Bacillus larvae*	i) American foul brood
(b) *Mellissococcus pluton*	ii) European foul brood
(c) *Aspergillus flavus*	iii) Sac brood
(d) RNA Virus	iv) Stone brood

4. Match the scientific name of natural enemies of honeybee

(a) *Merops orientalis*	i) Green bee eater
(b) *Vespa orientalis*	ii) Bee hunter wasp
(c) *Galleria melanoluca*	iii) Greater wax moth
(d) *Achroia grisella*	iv) Lesser wax moth

5. Match the causal organism and disease of honey bee

(a) *Malpighamobea mellificae*	i) Amoebic disease
(b) *Acarapis woodi*	ii) Isle of wight
(c) *Ascophaera apis*	iii) Chalk brood
(d) *Aspergillus flavus*	iv) Stone brood

6. **Match the causal organism and place of infection**

(a)	*Malpighamobea mellificae*	i)	Trachea
(b)	*Acarapis woodi*	ii)	Malpighian tubules
(c)	*Nosema apis*	iii)	Alimentary canal
(d)	*Aspergillus flavus*	iv)	Stomach

7. **Match the composition of propolis**

(a)	Resinous compound	i)	55 per cent
(b)	Bee wax	ii)	30 per cent
(c)	Aromatic oils	iii)	10 per cent
(d)	Pollen	iv)	5 per cent

8. **Match the products and their source of production in honey bee**

(a)	Royal jelly	i)	Poison gland
(b)	Queen substance	ii)	Mandibular gland
(c)	Bee venom	iii)	Hypopharyngeal gland
(d)	Bee wax	iv)	Wax gland

Answers

Multiple Choice Questions

1	(b)	21	(a)	41	(b)	61	(d)
2	(d)	22	(a)	42	(b)	62	(c)
3	(c)	23	(b)	43	(b)	63	(d)
4	(d)	24	(a)	44	(b)	64	(a)
5	(a)	25	(a)	45	(c)	65	(b)
6	(a)	26	(b)	46	(d)	66	(a)
7	(d)	27	(c)	47	(d)	67	(a)
8	(b)	28	(a)	48	(a)	68	(a)
9	(c)	29	(a)	49	(b)	69	(a)
10	(d)	30	(c)	50	(a)	70	(b)
11	(b)	31	(b)	51	(b)	71	(c)
12	(b)	32	(a)	52	(d)	72	(d)
13	(d)	33	(b)	53	(c)	73	(a)
14	(c)	34	(b)	54	(b)	74	(a)
15	(a)	35	(b)	55	(a)	75	(c)
16	(a)	36	(b)	56	(a)	76	(b)
17	(c)	37	(d)	57	(b)	77	(c)
18	(b)	38	(c)	58	(d)	78	(a)
19	(c)	39	(b)	59	(a)	79	(c)
20	(a)	40	(a)	60	(a)	80	(d)

81	(c)	91	(c)	101	(c)	111	(c)
82	(c)	92	(b)	102	(d)	112	(c)
83	(c)	93	(a)	103	(d)	113	(c)
84	(a)	94	(a)	104	(a)	114	(a)
85	(a)	95	(b)	105	(d)	115	(b)
86	(a)	96	(b)	106	(b)	116	(b)
87	(a)	97	(d)	107	(b)	117	(a)
88	(a)	98	(d)	108	(d)	118	(a)
89	(a)	99	(a)	109	(c)	119	(c)
90	(d)	100	(b)	110	(c)		

Fill in the Blanks

1. Pinocembrin
2. Apiary
3. Nectar
4. Five
5. 10 °C
6. Bee glue
7. Bee bread
8. Skin
9. Larval
10. Nuptial flight

Match the Following

1. a - iv, b - iii, c - ii, d - i
2. a - ii, b - iv, c - iii, d - i
3. a - i, b - ii, c - iv, d - iii
4. a - i, b - ii, c - iii, d - iv
5. a - i, b - ii, c - iii, d - iv
6. a - ii, b - i, c - iv, d - iii
7. a - i, b - ii, c - iii, d - iv
8. a - iii, b - ii, c - i, d - iv

Chapter 15
Silkworms and Sericulture

Multiple Choice Questions

1. Rearing and breeding of the silk moth to obtained the raw silk fibre is known as

(a) Silkworm rearing (b) Grainage
(c) Sericulture (d) Silk reeling

2. Sericin is a water soluble gelatinous protein forming which layer of the silk threads?

(a) Outer layer (b) Sub-surface layer
(c) Inner core (d) Middle layer

3. Silk is a natural fibre from the secretion by

(a) Silkworm (b) Spider
(c) Aphid (d) Both *(a)* & *(b)*

4. Fibroin produce which layer of the silk threads

(a) Outer layer (b) Sub-surface layer
(c) Inner core (d) Middle layer

5. Silk worm was discovered in

(a) 2640 BC (b) 640 AD
(c) 1640 BC (d) 1640 AD

6. Of the total silk that is produce in India mulberry silk accounts to

(a) 90 per cent (b) 80 per cent
(c) 75 per cent (d) 70 per cent

7. First country to know about silk is

(*a*) India (*b*) China

(*c*) Japan (*d*) France

8. A single cocoon constructed by two larvae is known as

(*a*) Single cocoon (*b*) Paired cocoon

(*c*) Double cocoon (*d*) Joint cocoon

9. Silk filament reeled out of double cocoon is known as

(*a*) Span Silk (*b*) Reeled silk

(*c*) Dupion silk (*d*) Raw silk

10. The mulberry silkworm belongs to the family

(*a*) Bombydae (*b*) Bombycidae

(*c*) Papilionidae (*d*) Saturnidae

11. Non-mulberry silkworms belongs to the family

(*a*) Bombydae (*b*) Bombycidae

(*c*) Papilionidae (*d*) Saturnidae

12. The newly hatched silkworms are known as

(*a*) Pharrate (*b*) Nymph

(*c*) Naiads (*d*) Chawki

13. Highest silk producing state in India is

(*a*) Uttar Pradesh (*b*) Assam

(*c*) Karnataka (*d*) West Bengal

14. Silkworm larvae parasitized by an endoparasite

(*a*) Uzi fly (*b*) Wasp

(*c*) Fruit fly (*d*) House fly

15. *Exorista sorbillans* is the scientific name of

(*a*) Uzi fly (*b*) Wasp

(*c*) Fruit fly (*d*) House fly

16. Semi domesticated species of silk worm is

(*a*) Mulberry (*b*) Muga

(*c*) Tasar (*d*) Eri

17. The silkworm species produce non reelable cocoon is

(*a*) Mulberry (*b*) Muga

(*c*) Tasar (*d*) Eri

18. The species of silkworm that monopoly in India is

(*a*) Mulberry (*b*) Muga

(*c*) Eri (*d*) Tasar

19. The giant silkworm species is

(*a*) *Attacus atlus* (*b*) *Antheraea assama*

(*c*) *Semia cynthia ricini* (*d*) *Bombyx mori*

20. The stage of mulberry silkworm that undergoes diapause is

(*a*) Egg (*b*) Larva

(*c*) Pupa (*d*) Adult

21. The silkworm inhabiting in subtropical regions (India) lay which type of eggs?

(*a*) Non-diapause (*b*) Diapause

(*c*) Unfertilized eggs (*d*) Only male eggs

22. The number of generations of an insect completes in one year in its natural condition is known as

(*a*) Moltinism (*b*) Voltinism

(*c*) Life cycle (*d*) Biology

23. The species of silkworm, where the cocoon has a peduncle

(*a*) Mulberry (*b*) Muga

(*c*) Nistari (*d*) Tasar

24. The silk spinning organ of silkworm is called as

(*a*) Needle (*b*) Spiting organ

(*c*) Spinneret (*d*) Silk gland

25. The golden yellow colour silk is produced by which silkworm

(*a*) Mulberry (*b*) Muga

(*c*) Nistari (*d*) Tasar

26. The cultivation of food plants of non mulberry silkworm is known as

(*a*) Moriculture (*b*) Silviculture

(*c*) Gardening (*d*) Arboriculture

27. The cultivation of mulberry plants is called

(*a*) Moriculture (*b*) Silviculture

(*c*) Gardening (*d*) Arboriculture

28. The unit used to measure the thickness of silk filaments is

(*a*) Diameter (*b*) Pie

(*c*) Centimetre (*d*) Denier

29. Denier measured as the weight of long silk filaments of

(a) 900 mt (b) 9000 mt

(c) 9900 mt (d) 10000 mt

30. The place where DFLs of silkworm is produced on scientific basis is known as

(a) Egg floor (b) Breeding chamber

(c) Matting box (d) Gainage hall

31. The stage of tasar silkworm that undergoes diapause is

(a) Egg (b) Larva

(c) Pupa (d) Adult

32. The common treatment to silkworm eggs to overcome hibernation is use of

(a) Hot H_2SO_4 (b) Hot HCl

(c) Cold H_2SO_4 (d) Cold HCl

33. The process of killing pupa without changing the cocoon shell layer is known as

(a) Nipping (b) Stiffing

(c) Ginning (d) Reeling

34. The stages of silkworm when it is synthesizing silk

(a) Pupa (b) 3rd inster larva

(c) 4th and 5th instar larva (d) 2nd inster larva

35. DFL stands for

(a) Demand for laying (b) Disease free laying

(c) Disease free larva (d) Disease full larva

36. Muga silkworm is a/an

(a) Univoltine (b) Bivoltine

(c) Multivoltie (d) Trivoltine

37. Fibroin made up of glycine, tyrosine and

(a) Seline (b) Alanine

(c) Leucine (d) Valine

38. Sericin made up of seline, leucine and

(a) Glycine (b) Alanine

(c) Tyrosine (d) Valine

39. In silkworms silk is produced by the

(*a*) Mandibuler gland (*b*) Pharyngial gland

(*c*) Labial gland (*d*) Maxilary gland

40. Most of the part of silk consists of

(*a*) Sericin (*b*) Fibroin

(*c*) Seline (*d*) Leucine

41. In general a single caterpillar of silkworm is capable to produce how much silkworm

(*a*) 650 - 1300 metres (*b*) 1650 - 1700 metres

(*c*) 100 - 650 metres (*d*) 1300 - 1650 metres

42. Totally domesticated silkworm is

(*a*) Mulberry (*b*) Muga

(*c*) Tasar (*d*) Oak tasar

43. How many mulberry cuttings are required for one hectare land if 2 cuttings are planted in each pit?

(*a*) 100000 (*b*) 10000

(*c*) 20000 (*d*) 2000

44. In sericulture, "Chandraki" a basket with a spiral wall, made up of bamboo slips is use as

(*a*) Coupling Box (*b*) Mounting

(*c*) Mountage (*d*) Knife

45. To obtain good quality silk which type of leaves are required for larval feeding

(*a*) Yellow leaves (*b*) Matured leaves

(*c*) Diseased leaves (*d*) Grownup or tender leaves

46. One DFL of mulberry silkworm consist of how many eggs

(*a*) 50 - 100 (*b*) 100 - 200

(*c*) 200 - 300 (*d*) 300 - 400

47. Which set of the following are the primary host plants of muga silkworm?

(*a*) Mulberry and Sal (*b*) Arjun and Asan

(*c*) Castor and Ber (*d*) Som and Soalu

48. Which one of the following is Oak tasar silkworm?

(*a*) *Semia cynthia ricini* (*b*) *Antheraea assama*

(*c*) *Bombyx mori* (*d*) *Antheraea prolei*

49. The second largest producer of silk in world is

(*a*) China (*b*) India

(*c*) Japan (*d*) France

50. A compact cocoon of a silkworm contains

(*a*) Pupa (*b*) Last larval Skin

(*c*) Larva (*d*) Both (*a*) & (*b*)

51. Which type of antennae muga silkworm have?

(*a*) Bipectinate (*b*) Aristate

(*c*) Plumose (*d*) Geniculate

52. Muga silkworm is a

(*a*) Monophagous (*b*) Oligophagous

(*c*) Polyphagous (*d*) Stenophagous

53. Mulberry silkworm is a

(*a*) Monophagous (*b*) Oligophagous

(*c*) Polyphagous (*d*) Sap sucker

54. Anal horns are found in the larva of

(*a*) *Semia cynthia ricini* (*b*) *Antheraea assama*

(*c*) *Bombyx mori* (*d*) *Antherae prolei*

55. Central Silk Research Station is located at

(*a*) West Bengal (*b*) Rajasthan

(*c*) Uttar Pradesh (*d*) Maharashtra

56. Green silkworm is scientifically known as

(*a*) *Attacus ricini* (*b*) *Antheraea assamensis*

(*c*) *Antheraea paphia* (*d*) *Antheraea mylitta*

57. *Attacus ricini* originated from

(*a*) Bihar (*b*) Karnataka

(*c*) Assam (*d*) Kashmir

58. Central Silk Board established in the year

(*a*) 1949 (*b*) 1948

(*c*) 1947 (*d*) 1946

59. Chemically silk thread are

(*a*) Carbohydrate (*b*) Protein

(*c*) Fat (*d*) Vitamin

60. Muga, Tasar and Eri silkworm belongs to the family

(*a*) Bombycidae (*b*) Papilionidae

(*c*) Pyralidae (*d*) Saturniidae

61. In the production of mulberry silk India's place in the world is

(*a*) First (*b*) Second

(*c*) Third (*d*) Fourth

62. The fertilized female of mulberry silkworm lays eggs within 24 hours ranging from

(*a*) Less than 100 (*b*) 100 - 200

(*c*) 200 - 300 (*d*) 400 - 500

63. Silk obtained from sources other than mulberry are generally termed as

(*a*) Tasar (*b*) Muga

(*c*) Non mulberry (*d*) Mulberry

64. Which country is the leading producer of oak tasar (*Antheraea pernyi*) in the world?

(*a*) India (*b*) China

(*c*) Japan (*d*) Nepal

65. Silk spun formed open ended cocoons in case of

(*a*) Mulberry silkworm (*b*) Tasar silkworm

(*c*) Muga silkworm (*d*) Eri silkworm

66. Which silk now has secured geographical indicators status under Intellectual Property Regime (IPR)?

(*a*) Mulberry silk (*b*) Tasar silk

(*c*) Muga silk (*d*) Eri silk

67. The silk producing apparatus is a paired organ consisting of modified

(*a*) Mandibular gland (*b*) Maxillary gland

(*c*) Labial salivary gland (*d*) Labral gland

68. The common method of propagation of mulberry plants is by

(*a*) Seed (*b*) Stem cutting

(*c*) Tissue culture (*d*) Root cutting

69. Approximate weight of a single mulberry cocoon is

(*a*) 1.25 grams (*b*) 1.67 grams

(*c*) 2.02 grams (*d*) 2.33 grams

70. To produce 1 kg of mulberry silk how much of green foliage is required

(*a*) 50 - 70 kg (*b*) 80 - 100 kg

(*c*) 120 - 160 kg (*d*) 180 - 200 kg

71. Number of kg of cocoons required to produce 1 kg of raw silk yarn is measured in

(*a*) Bale (*b*) Denier

(*c*) Renditta (*d*) None of these

72. The process of transfer of mature fifth inster larvae to mountage is called

(*a*) Mounting (*b*) Reeling

(*c*) Curing (*d*) Stiffing

73. Muscardine disease in silk worms is caused by

(*a*) *Metharizium* spp (*b*) *Nosema bombysis*

(*c*) *Bacillus thuringiensis* (*d*) *Beauveria basiana*

74. Banya silk are came from

(*a*) Mulberry silkworm (*b*) Wild silkworm

(*c*) Tasar silkworm (*d*) Muga silkworm

75. The origin of silk glands is

(*a*) Ectoderm (*b*) Mesoderm

(*c*) Endoderm (*d*) None of these

76. The only country the producers all the five commercial types of silk is

(*a*) India (*b*) USA

(*c*) China (*d*) Japan

77. The disease in silkworms which is characterized by the emission of foul smell is

(*a*) Muscardine (*b*) Pebrine

(*c*) Flacherie (*d*) Grasserie

78. The main host plant of Eri-silkworm is

(*a*) Mulberry (*b*) Acacia

(*c*) Terminalia (*d*) Castor

79. Disinfestations of silkworm eggs is done with

(*a*) 2 per cent Formalin (*b*) 5 per cent Formalin

(*c*) 20 per cent Formalin (*d*) None of these

80. The newly hatched silkworms are called as

(*a*) Kegs (*b*) Ants

(*c*) Chawki (*d*) All of these

81. Muscardine disease in silkworms is caused by

(*a*) Fungus (*b*) Bacteria

(*c*) NPV (*d*) Sporozoan

82. *Nosema bombysis* causes a disease in silkworms is known as

(*a*) Muscardine (*b*) Pebrine

(*c*) Flacherie (*d*) Grasserie

83. The disease in silkworms in which infection spreads to successive generations through eggs of a diseased moth is

(*a*) Muscardine (*b*) Pebrine

(*c*) Flacherie (*d*) Grasserie

84. Silkworms are very sensitive to the following bio-pesticides

(*a*) *Bacillus thuringiensis* (*b*) *Metarrhizium* sp.

(*c*) *Verticilium* sp. (*d*) *Nomurea rileyi*

85. Seri-biotech Research Laboratory are situated as

(*a*) Bangaluru (*b*) Mysore

(*c*) Kolkata (*d*) Goa

86. The silkworm that is reared in indoors

(*a*) Mulberry (*b*) Eri

(*c*) Tasar (*d*) Both (*a*) & (*b*)

87. The silkworm that is reared outdoors (Natural conditions)

(*a*) Tasar (*b*) Muga

(*c*) Mulberry (*d*) Both (*a*) & (*b*)

88. To produce 1 kg of raw silk, the quantity of cocoons required is

(*a*) 7 to 8 kg (*b*) 8 to 9 kg

(*c*) 9 to 10 kg (*d*) 10 to 11 kg

89. Indian sericulture is mostly

(*a*) Univoltine (*b*) Bivoltine

(*c*) Multivoltine (*d*) None of these

90. Silk is produced by__________stage of silkworm.

(*a*) Egg (*b*) End of larval

(*c*) Pupal (*d*) Adult

91. A female eri silkworm moth lays approximately__________numbers of eggs.

(*a*) 100 - 200 (*b*) 200 - 300

(*c*) 300 - 500 (*d*) 500 - 600

92. The hook like structure present at the base of the prolegs of silkworm larvae are called

(*a*) Crochets (*b*) Claw

(*c*) Propagules (*d*) Hooks

93. The amino acid which is not abundant in silk fibroin

(*a*) Alamine (*b*) Glycine

(*c*) Serine (*d*) Thamine

94. Ninety percent (90 per cent) of the total silk produced in the country is from

(*a*) Univoltine (*b*) Bivoltine

(*c*) Multivoltine (*d*) Minivoltine

95. The variety of mulberry that has been reportedly to yield maximum (60 MT) in the world

(*a*) V 1 (*b*) S 36

(*c*) MR 2 (*d*) K 2

Fill in the Blanks

1. The oak tasar silkworm is scientifically known as__________.

2. __________is a valuable natural protein fiber produced by certain insects.

3. The largest producer of eri silk in the world is__________.

4. The headquarter of Indian Silk Export Promotion Council (ISEPC) located at__________.

5. Numbers of Chromosome found in Mulberry silkworm, *Bombyx mori* is__________.

6. The first evidence of the silk trade is an Egyptian mummy of__________BC.

7. The culture of silkworms (*Bombyx mori*) began in China about__________years ago.

8. __________is a large factory where reeling of cocoons is carried out with the help of advanced technology.

9. Newly hatched silkworm larva is technology known as__________.

10. Two larvae spin a single cocoon, which are unsuitable for reeling purpose, is called as__________.

11. Just after the harvesting, the cocoon having live pupa inside, is called as__________.

12. Post cocoon processing includes stiffing and__________.
13. In India__________state is the leading producer of mulberry silk.
14. In India__________state is the leading producer of non-mulberry silk.
15. India has monopoly in the production of__________silk.
16. Muga silkworm belongs to the family__________.
17. The largest producer of silk in the world is__________.
18. The largest consumer of silk in the world is__________.
19. The silk filament is composed of two fibroin filaments held together by a cementing layer of__________.
20. Fibroin and Sericin account for about__________and__________per cent of the raw silk by weight respectively.
21. __________disease of silkworms is manage only through mother moth examination.
22. In India__________state is the leading producer of Muga silk.
23. An Indian silk bale is of__________kg.
24. __________type of antenna present in silkworm.
25. The killing of pupae of silkworm in the cocoon is called__________.
26. The process of removing the threads from killed cocoons is called as__________.
27. __________disease of silkworm is caused by Nuclear Polyhedrosis Virus.
28. __________disease of silkworm is caused by bacteria.

Match the Following

1. **Match the following disease with their group of pathogens**

(a) Flacherie	i) Virus
(b) Grasserie	ii) Bacteria
(c) Muscardine	iii) Fungus
(d) Pebrine	iv) Protozoa

2. Match the following disease of silkworm and their causal organism

(a) Flacherie	i) *Bacillus bombycis*
(b) Grasserie	ii) *Nosema bombysis*
(c) Muscardine	iii) BmNPV
(d) Pebrine	iv) *Beuveria bassiana*

3. Match the following silkworm with scientific name

(a) Mulberry silkworm	i) *Antheraea assama*
(b) Eri silkworm	ii) *Semia cynthia ricini*
(c) Tasar silkworm	iii) *Antheraea mylitta*
(d) Muga silkworm	iv) *Bombyx mori*

4. Match the following silkworm with their food plant

(a) Mulberry silkworm	i) *Ricinus communis*
(b) Eri silkworm	ii) *Terminalia tomentasa*
(c) Tasar silkworm	iii) *Moras alba*
(d) Muga silkworm	iv) *Persia bombycina*

5. Match the following silkworm with their food plant

(a) *Bombyx mori*	i) Mulberry
(b) *Semia cynthia ricini*	ii) Castor
(c) *Antheraea mylitta*	iii) Assan
(d) *Antheraea assama*	iv) Som

6. Match the following institute and their location

(a) Central Silk Board (CSB)	i) Ranchi
(b) Central Tasar Research and Training Institute	ii) Bangalore
(c) Central Muga Eri Research and Training Institute	iii) Mysore
(d) Central Sericulture Research and Training Institutes (CSR and TI)	iv) Ladoigarh

7. Match the type of silk with their colour

(a) Nistari	i) Golden yellow
(b) Muga silk	ii) Yellow
(c) Tasar silk	iii) Brown
(d) Eri silk	iv) White or Brick-red

Answers

Multiple Choice Questions

1	(c)	21	(a)	41	(a)	61	(b)
2	(a)	22	(b)	42	(a)	62	(d)
3	(d)	23	(d)	43	(c)	63	(c)
4	(c)	24	(c)	44	(c)	64	(b)
5	(a)	25	(b)	45	(d)	65	(d)
6	(a)	26	(d)	46	(d)	66	(c)
7	(b)	27	(a)	47	(d)	67	(c)
8	(c)	28	(d)	48	(d)	68	(b)
9	(c)	29	(b)	49	(b)	69	(b)
10	(b)	30	(d)	50	(d)	70	(c)
11	(d)	31	(c)	51	(a)	71	(c)
12	(d)	32	(b)	52	(c)	72	(a)
13	(c)	33	(b)	53	(a)	73	(d)
14	(a)	34	(c)	54	(c)	74	(b)
15	(a)	35	(b)	55	(a)	75	(a)
16	(b)	36	(c)	56	(b)	76	(a)
17	(d)	37	(c)	57	(c)	77	(c)
18	(b)	38	(a)	58	(a)	78	(d)
19	(a)	39	(c)	59	(b)	79	(a)
20	(a)	40	(b)	60	(d)	80	(d)

81	(a)	85	(a)	89	(c)	93	(d)
82	(b)	86	(d)	90	(b)	94	(c)
83	(b)	87	(d)	91	(c)	95	(a)
84	(a)	88	(a)	92	(a)		

Fill in the Blanks

1. *Antheraea prolei*
2. Silk
3. India
4. Mumbai
5. n = 28
6. 1070 BC
7. 5000
8. Filature
9. Kengo
10. Double cocoon
11. Green cocoon
12. Reeling
13. Karnataka
14. Assam
15. Muga
16. Saturniidae
17. China
18. India
19. Sericin
20. 75 and 25
21. Pebrine
22. Assam
23. 50 kg
24. Bipectinate
25. Stiffing

26. **Reeling**

27. **Grasserie**

28. **Flacherie**

Match the Following

1. a) - ii, b) - i, c) - iii, d) - iv
2. a) - i, b) - iii, c) - iv, d) - ii
3. a) - iv, b) - ii, c) - iii, d) - i
4. a) - iii, b) - i, c) - ii, d) - iv
5. a) - i, b) - ii, c) - iii, d) - iv
6. a) - ii, b) - i, c) - iv, d) - iii
7. a) - ii, b) - i, c) - iii, d) - iv

Chapter 16
Lac Insect

Multiple Choice Questions

1. **The Kusumi strain is raised mainly on__________and on other host with kusum brood.**
   (*a*) Kusum (*b*) Ber
   (*c*) Shimul (*d*) Palas
2. **The parasitiod of lac insect belong to the superfamily**
   (*a*) Chalcidoidea (*b*) Tettigoniodea
   (*c*) Vespidae (*d*) Apoidea
3. **Lac insects inoculated through**
   (*a*) Seed lac (*b*) Brood lac
   (*c*) Aric lac (*d*) Dust lac
4. **The appearance of an__________in the vicinity of the anal tubercle is the accurate indication of the mature crop.**
   (*a*) Pink area (*b*) Green area
   (*c*) White area (*d*) Orange yellow area
5. **The nymph loses its eyes, legs and antennae after**
   (*a*) First moult (*b*) Second moult
   (*c*) Third moult (*d*) Fourth moult
6. **The melting seed lac produced a button shaped pieces are known as**
   (*a*) Brood lac (*b*) Stick lac
   (*c*) Seed lac (*d*) Shellac

7. Escape the nymphs from the brood lac, what has been left is the

(*a*) Brood lac (*b*) Stick lac

(*c*) Seed lac (*d*) Shellac

8. Lac insect belongs to the order

(*a*) Hymenoptera (*b*) Homoptera

(*c*) Diptera (*d*) Isoptera

9. Which of the following is not a host of lac insects

(*a*) Kusum (*b*) Mulberry

(*c*) Palas (*d*) Ber

10. In propagation of Lac insect, brood lac sticks or twigs should be cut when ovisac turns

(*a*) Pink (*b*) Green

(*c*) White (*d*) Half orange and half red

11. How many generation of lac insect in a year

(*a*) One (*b*) Two

(*c*) Three (*d*) Four

12. The newly hatched (crawler) of lac insect is__________in colour.

(*a*) Red (*b*) Pinkish

(*c*) Black (*d*) Green

13. India is the major producer of lac accounting for about__________of the total world lac production.

(*a*) 70 per cent (*b*) 50 per cent

(*c*) 55 per cent (*d*) 60 per cent

14. In India the most common host trees for Lac production are

(*a*) Palas (*Butea monosperma*) (*b*) Ber (*Zizyphus jujube*)

(*c*) Kusum (*Schleichera oleosa*) (*d*) All of these

15. Lac is a resinous secretion by an insect

(*a*) *Bombyx mori* (*b*) *Apis melifera*

(*c*) *Laccifer lacca* (*d*) *Lycosa* Sp.

16. *Laccifer lacca* Ker belongs to the family

(*a*) Apidae (*b*) Lacciferidae

(*c*) Saturnidae (*d*) Tetranicidae

17. The only resin of animal origin is

(*a*) Wax (*b*) Silk
(*c*) Lac (*d*) Honey

18. The type of reproduction in Lac insect is

(*a*) Ovoviviparous (*b*) Viviparous
(*c*) Oviparous (*d*) None of these

19. From both fertilized and unfertilized egg laid by female of lac insect __________nymphs are hatched out.

(*a*) Only Male (*b*) Only Female
(*c*) Both Male and Female (*d*) None of these

20. The lac gland is situated at__________the cuticle all over the body of nymph.

(*a*) Under (*b*) Above
(*c*) Over (*d*) Between

21. The share of rangeeni strain of lac insect to lac production in India is

(*a*) 80-90 per cent (*b*) 90-100 per cent
(*c*) 70-80 per cent (*d*) 60-70 per cent

22. The lac insect moult__________before reaching maturity.

(*a*) Once (*b*) Twice
(*c*) Thrice (*d*) Four times

23. The glands that help in secretion of Lac

(*a*) Resin glands (*b*) Salivary glands
(*c*) Pharyngeal glands (*d*) Endocrine glands

24. The chief producer of lac are

(*a*) Female (*b*) Male
(*c*) Egg (*d*) None of these

25. There are two distinct strain of lac insect in India are

(*a*) Kusumi and Rangeeni (*b*) Katki and Jethui
(*c*) Kotia and Bhodia (*d*) Kusumi and Jethui

26. The leading producer of Lac in the world is

(*a*) India (*b*) China
(*c*) Thailand (*d*) Bangladesh

27. Lac sticks bearing alive mother cells which produce lac larvae are known as

(*a*) Brood lac (*b*) Stick lac
(*c*) Seed lac (*d*) Shellac

28. The major predator of Lac insect is

(*a*) *Coccinella septempunctata* (*b*) *Eublemma amabilis*

(*c*) *Chrysopa* spp. (*d*) All of these

29. The scraped lac from sticks are crushed lightly, washed in water to remove dead insect, dry matter is known as

(*a*) Brood lac (*b*) Stick lac

(*c*) Seed lac (*d*) Shellac

30. The resin glands are distributed all over the body except

(*a*) Mouth parts (*b*) Two breathing pores

(*c*) Anus (*d*) All of these

31. The highest Lac producing state of India is

(*a*) Chattisgarh (*b*) Bihar

(*c*) Jharkhand (*d*) Madhya Pradesh

32. Indian Institute of Natural Resins and Gums (IINRG) is situated at

(*a*) Bhopal (*b*) Raipur

(*c*) Ranchi (*d*) Kolkata

33. Lac insect belongs to the family

(*a*) Gerridae (*b*) Lacciferridae

(*c*) Fulgoridae (*d*) Coccidae

34. Rangeeni strain is related to

(*a*) Silkworm (*b*) Honey bee

(*c*) Lac insect (*d*) None of these

35. The Indian Lac Research Institute was established in

(*a*) 1923 (*b*) 1924

(*c*) 1925 (*d*) 1926

36. Indian Lac Research Institute is located at

(*a*) Madhya Pradesh (*b*) Bihar

(*c*) Jharkhand (*d*) Chhattisgarh

37. The percentage of resin in lac is

(*a*) 68 - 90 per cent (*b*) 92 - 95 per cent

(*c*) 60 - 65 per cent (*d*) 58 - 60 per cent

38. The share of following crop of Lac is highest in India

(*a*) Katki (*b*) Baishaki

(*c*) Aghani (*d*) Jethwi

39. India export shellac mainly to

(a) Japan (b) Korea

(c) China (d) USA

40. The strain of the lac insect which survives on all hosts is known as

(a) Kusumi (b) Rangini

(c) Lakhini (d) Namkumi

41. The fertilized female of *Laccifer lacca* lays eggs ranging from

(a) Less than 100 (b) 100 - 200

(c) 200 - 500 (d) 500 - 700

42. The strain of lac insect that is reared on all hosts of lac except kusum tree is known as

(a) Kusumi (b) Rangini

(c) Lakhini (d) Namkumi

43. The emergence of nymphs of lac insect is called

(a) Swarming (b) Crawling

(c) Teeming (d) Heaving

44. Lac is usually marketed as

(a) Brood lac (b) Stick lac

(c) Seed lac (d) Shellac

45. The cocoon which is the raw material for shellac and is called

(a) Brood lac (b) Stick lac

(c) Seed lac (d) None of these

46. The host plant of Kusumi strain *is*

(a) *Butea monosperma* (b) *Zizyphus jujube*

(c) *Schleichera oleosa* (d) *Ficus religiosa*

47. Second largest lac producing state of India is

(a) Jharkhand (b) West Bengal

(c) Assam (d) Madhya Pradesh

Fill in the Blanks

1. Immature harvesting where lac is harvested before swarming and this type of lac is called__________.

2. Ari lac gives better production in__________crop.

3. In lac culture, all the trees of host plants of a definite area are artificially divided or marked in five groups which is known as__________.

4. The host plant twig bearing the lac along with eggs is called__________stick.

Match the Following

1. Match the following trees with their scientific name

(a)	Palas	i)	*Schleichera oleosa*
(b)	Kusum	ii)	*Butea monosperma*
(c)	Ber	iii)	*Bombyx ciba*
(d)	Shimul	iv)	*Zizyphus jujube*

2. Match the trees with their contribution in lac production

(a)	Palas	i)	4 per cent
(b)	Kusum	ii)	10 per cent
(c)	Ber	iii)	33 per cent
(d)	Others	iv)	53 per cent

3. Match the share of different crops in lac production in India

(a)	Baisakhi	i)	27.40 per cent
(b)	Katki	ii)	15.85 per cent
(c)	Aghani	iii)	38.66 per cent
(d)	Jethui	iv)	18.09 per cent

4. Match the following lac crops with their time of harvest

(a)	Aghani	i)	May - June
(b)	Jethui	ii)	January - February
(c)	Katki	iii)	June - July
(d)	Baisakhi	iv)	October - November

Answers

Multiple Choice Questions

1	(a)	13	(a)	25	(a)	37	(a)
2	(a)	14	(d)	26	(a)	38	(b)
3	(b)	15	(c)	27	(a)	39	(d)
4	(d)	16	(b)	28	(b)	40	(b)
5	(a)	17	(c)	29	(c)	41	(c)
6	(d)	18	(a)	30	(d)	42	(b)
7	(b)	19	(c)	31	(a)	43	(a)
8	(b)	20	(a)	32	(c)	44	(d)
9	(b)	21	(a)	33	(b)	45	(b)
10	(a)	22	(c)	34	(c)	46	(c)
11	(b)	23	(a)	35	(c)	47	(a)
12	(b)	24	(a)	36	(c)		

Fill in the Blanks

1. Ari lac
2. Rangini
3. Ceupe system
4. Brood lac

Match the Following

1. a) - ii, b) - i, c) - iv, d) - iii
2. a) - iv, b) - iii, c) - ii, d) - i
3. a) - iii, b) - i, c) - iv, d) - ii
4. a) - ii, b) - iii, c) - iv, d) - i

Chapter 17

Other Useful Insects

Multiple Choice Questions

1. **Which of the following are helpful insects?**
   (a) Leaf miner (b) Vector
   (c) Predators (d) None of these

2. **A scale insect, *Ericerus pela* is used in China for commercial production of**
   (a) Wax (b) Honey
   (c) Silk (d) Lac (Resin)

3. **Which stage of *Ericerus pela* produced economically valuable product?**
   (a) 2nd instar male (b) 2nd instar female
   (c) 3rd inster male (d) 1st inster female

4. **Bumble bees are more effective pollinators of**
   (a) Large sized flowers (b) Small sized flowers
   (c) Medium sized flowers (d) None of these

5. ***Cactoblastis cactorum* (Pyralidae : Lepidoptera) were introduced to control a cactus *Opuntia* sp. from Argentina to**
   (a) Australia (b) India
   (c) Japan (d) England

6. **The scarlet dye (cochineal dye) producing insect is a scale insect *Dactylopius ceylonicus* feed on**
   (a) *Lantena* sp. (b) *Opuntia* sp.
   (c) *Cynodon* sp. (d) *Chenopodium* sp.

7. In India *Opuntia dillenii* was eradicated by *Dactylopius tomentosus* introduced from Maxico via

(*a*) Sri Lanka (*b*) Nepal

(*c*) China (*d*) Bhutan

8. The insect species, *Teleonemia scrupulosa* feeds on *Lantana camara* is a

(*a*) Lace bug (*b*) Seed fly

(*c*) Seed midge (*d*) Beeetle

9. The scientific name of seed fly feeds on *Lantana camara* is a

(*a*) *Ophiomuia lantanae* (*b*) *Teleonemia scrupulosa*

(*c*) *Opuntia dillenii* (*d*) *Chrysolina* spp.

10. St. John's Wart or Klamath (*Hypericum perforatum*) is controlled by a leaf feeding beetle

(*a*) *Ophiomuia lantanae* (*b*) *Teleonemia scrupulosa*

(*c*) *Opuntia dillenii* (*d*) *Chrysolina* spp.

11. Major group of scavengers are

(*a*) Coleoptera and Diptera

(*b*) Hymenoptera and Lepidoptera

(*c*) Odonota and Hymenoptera

(*d*) None of these

12. In advance genetic research insect use are

(*a*) *Dosophila melanogaster* (*b*) *Antheraea assama*

(*c*) *Scirpophaga incertulas* (*d*) *Musca domestica*

13. Fruit flies, *Drosophila melanogaster* have how many pairs of chromosome, only

(*a*) Four pairs (*b*) Five pairs

(*c*) Six pairs (*d*) Seven pairs

14. Butterflies and dragonflies are good indicator of

(*a*) Hailstorm (*b*) Depression

(*c*) Climatic changes (*d*) Cyclone

15. Cochineal dye is obtained from dried bodies of *Dactylopius coccus* and *Dactylopius tomentosus* are

(*a*) Scale insect (*b*) Beetle

(*c*) Moths (*d*) Ant

16. Cochineal contains__________percent carminic acid.

(*a*) 5 (*b*) 10

(*c*) 15 (*d*) 20

17. The stings of honey bees have remedial value in the treatment of

(*a*) Arthritis and Rheumatism (*b*) Kala-azar and Diphtheria

(*c*) Fever and Dengue (*d*) Diarrhoea

18. Cantharidin is a substance found in an insect, *Lytta vesicatoria* is a

(*a*) Blister beetle (*b*) Water beetle

(*c*) Water bug (*d*) Ground beetle

19. Cantharidin is useful internally treating certain

(*a*) Urinary disease (*b*) Diphtheria

(*c*) Diarrhoea (*d*) Kala-azar

20. Cantharidin is useful externally use as a

(*a*) Hair tonic (*b*) Shampoo

(*c*) Beauty cream (*d*) Hair dye

21. Cantharidin is now banded as medicine because of its

(*a*) High irritability (*b*) High toxicity

(*c*) High inflammation (*d*) High price

22. Medicine prepared from *Bllata americana* is used for

(*a*) Tiredness and urinary ailments

(*b*) Diphtheria and Sleeping sickness

(*c*) Diarrhoea

(*d*) Kala-azar

23. Medicine prepared from *Bllata orientalis* used for

(*a*) Urinary irritation, diphtheria etc. (*b*) Sleeping sickness

(*c*) Diarrhoea (*d*) Kala-azar

24. "Chinese medicine for eternal youth" is extracted from, *Cordyceps simensis* is a/an

(*a*) Ant (*b*) Beetle

(*c*) Caterpillars (*d*) Dragonfly

25. Allantoin is a substance isolated from the secretions of fly maggots which seems to have properties of healing

(*a*) Deep wound (*b*) Small wounds

(*c*) Little wounds (*d*) Shallow wounds

26. For "Maggot Therapy" dipteran larvae used from the family

(*a*) Oestridae (*b*) Calliphoridae

(*c*) Sarcophagidae (*d*) All of these

27. The most common species used for "Maggot Therapy" is

(*a*) Blue fly (*b*) Green stink midge

(*c*) Green peach aphid (*d*) Green bottle fly

28. Green bottle fly, *Lucilia sericata* is a major pest of sheep, causing __________disease, which may fatal.

(*a*) Fever (*b*) Strike

(*c*) Diarrhoea (*d*) Kala-azar

29. The eating of insects by man or other animals is known as

(*a*) Entomophyli (*b*) Entomophagy

(*c*) Insectofagy (*d*) Insectophyli

30. The male of giant water bug (*Lethocerus indicus*) is preferred for eating because of their unique flavour perhaps due to their

(*a*) Sex pheromone (*b*) Juvenile hormone

(*c*) Gonad secretion (*d*) Seminal fluid

31. Now a days for conservation of butterflies and beautification established many

(*a*) Butterfly sanctuaries (*b*) Butterfly gardens

(*c*) Butterfly parks (*d*) Butterfly Zone

32. In case of mango pollination is done by

(*a*) Mosquito (*b*) House fly

(*c*) Cockroach (*d*) Ant

33. Shiny metallic elytra of buprestid beetles are commonly used for

(*a*) Ornamentation (*b*) Soap

(*c*) Manure (*d*) Fish baits

34. Larvae of certain aquatic neuropteran used as

(*a*) Ornamentation (*b*) Soap

(*c*) Manure (*d*) Fish baits

35. The red carmine dye producing insects are belongs to which of the following group

(*a*) Moth (*b*) Coccinella

(*c*) Homoptera (*d*) Grasshopper

36. In scientific research the most well suited insect resource next only to *Drosophila* is

(*a*) Green bottle fly (*b*) Pollu beetle
(*c*) Silk worm (*d*) None of these

Fill in the Blanks

1. Ants, beetles, bees, larvae of flies, crickets, cutworms, collembolans etc. are lives in the__________

2. Insects those dig soil, add excreta, saliva, exuviae and dead bodies with soil are known as__________

3. "Specific Medicine Apis" is extracted from__________and used for medicinal purpose.

4. Practice of renting bee colonies for pollination service was first started in__________.

5. Some insects are very valuable because they kill insect pests of crops are called__________.

Match the Following

1. Match the following insects with their scientific name

(*a*) Buffalo fly i) *Drosophila melanogaster*
(*b*) Bush fly ii) *Musca domestica*
(*c*) Fruit fly iii) *Haematobiairritanis exigua*
(*d*) House fly iv) *Musca vetustissima*

2. Match the following coleopteran beetle with their family

(*a*) Rove beetle i) Tenebrionidae
(*b*) Skin beetle ii) Staphylinidae
(*c*) Carrion beetle iii) Dermestidae
(*d*) Darkling beetle iv) Silphidae

3. Match the following coleopteran beetles act as scavengers on following substance

(*a*) Chaffer beetle i) Wool, fur, hide etc.
(*b*) Rove beetle ii) Dead animals
(*c*) Skin beetle iii) Dung
(*d*) Carrion beetle iv) Decaying matter

4. Match the following dipteran flies with their family

(a) Fungus gnats	i) Tipulidae
(b) Dady-long-legs	ii) Psychodidae
(c) Sand fly	iii) Syrphidae
(d) Hover flies	iv) Mycetophilidae

5. Match the following dipteran flies act as scavengers

(a) Fungus gnats	i) Decaying organic matter
(b) Dady-long-legs	ii) Decaying vegetable matter
(c) Sand fly	iii) Organic matter
(d) Hover flies	iv) Decay of wood, fruit etc.

6. Match the following important insect pollinators with their scientific name

(a) Solitary bees	i) *Xylocopa* sp.
(b) Bumble bees	ii) *Bombus* sp.
(c) House fly	iii) *Musca* sp.
(d) Stingless bee	iv) *Trigona* sp.

Answers

Multiple Choice Questions

1	(c)	10	(d)	19	(a)	28	(b)
2	(a)	11	(a)	20	(a)	29	(b)
3	(a)	12	(a)	21	(b)	30	(a)
4	(a)	13	(a)	22	(a)	31	(c)
5	(a)	14	(c)	23	(a)	32	(b)
6	(b)	15	(a)	24	(c)	33	(a)
7	(a)	16	(b)	25	(a)	34	(d)
8	(a)	17	(a)	26	(d)	35	(c)
9	(a)	18	(a)	27	(d)	36	(c)

Fill in the Blanks

1. Soils
2. Soil builders
3. Honey bee
4. USA
5. Entomophagous insects

Match the Following

1. a) - iii, b) - iv, c) - i, d) - ii
2. a) - ii, b) - iii, c) - iv, d) - i
3. a) - iii, b) - iv, c) - i, d) - ii
4. a) - iv, b) - i, c) - ii, d) - iii
5. a) - iv, b) - ii, c) - iii, d) - i
6. a) - i, b) - ii, c) - iii, d) - iv

Chapter 18

Pests of Medical Importance

Multiple Choice Questions

1. **Leishmaniasis (*Kala azar*) disease caused by *Leishmania donovani* is a**
   - (*a*) Sand fly
   - (*b*) Protozoa
   - (*c*) House fly
   - (*d*) Bacteria
2. **Vector of river blindness (Onchocerciasis) disease is**
   - (*a*) *Anopheles* spp.
   - (*b*) *Glassina* spp.
   - (*c*) *Simulium* spp.
   - (*d*) *Musca* spp.
3. **Disease causing organism of *Kala-azar* is transmitted by**
   - (*a*) White fly
   - (*b*) Black fly
   - (*c*) Sand fly
   - (*d*) Tsetse fly
4. **Vector of Sleeping sickness disease is**
   - (*a*) *Anopheles* spp.
   - (*b*) *Glassina* spp.
   - (*c*) *Simulium* spp.
   - (*d*) *Musca* spp.
5. **Which one of the following is a vector of Chagas diseases?**
   - (*a*) *Rhodnius prolixus*
   - (*b*) *Blatta orientalis*
   - (*c*) *Culex pipiens*
   - (*d*) *Anophelus* spp.
6. **Causal organism of malaria in man is**
   - (*a*) *Plasmodium* spp.
   - (*b*) *Ricketsia mooseri*
   - (*c*) *Bacillus* spp
   - (*d*) None of these

7. Vector of amoebic dysentery disease is

(*a*) *Anopheles* spp. (*b*) *Glassina* spp.

(*c*) *Simulium* spp. (*d*) *Musca* spp.

8. Dengue disease is transmitted by

(*a*) *Anopheles* spp. (*b*) *Aedes* spp.

(*c*) *Culex* spp. (*d*) None of these

9. Which of the following disease is transmitted by human lice?

(*a*) Relapsing fever (*b*) Trench fever

(*c*) Epidemic typhus (*d*) All of these

10. Disease transmitted by sandflies

(*a*) Anthrax (*b*) *Kala-azar*

(*c*) Leishmaniasis (*d*) All of these

11. The most effective repellent of mosquitoes is

(*a*) Oil of citronella (*b*) Dimethyl phthalate

(*c*) Diethyl toluamide (DEET) (*d*) All of these

12. The insecticide most effective against mosquito is

(*a*) Deltamethrin (*b*) Cypermethrin

(*c*) Zinc Phosphide (*d*) Imidacloprid

13. Disease transmitted by eyeflies

(*a*) Ophthalmia (*b*) Conjunctivitis

(*c*) Both (*a*) & (*b*) (*d*) None of these

14. Which of the following is used as biological control agent in mosquito control?

(*a*) *Labio rohita.* (*b*) *Gambusia affinis*

(*c*) *Catla catla* (*d*) All of these

15. *Bovicola* spp. is a notorious pest of

(*a*) Man (*b*) Goat

(*c*) Pig (*d*) Sheep

16. Plague is transmitted through

(*a*) *Bemisia tabaci* (*b*) *Thrips tabaci*

(*c*) *Bovicola partellus* (*d*) *Xenopsylla cheopsis*

17. Blood sucking mosquito is

(*a*) Male (*b*) Female

(*c*) Both male and female (*d*) None of these

18. In India Malaria disease was first identified in

(*a*) Hyderabad (*b*) Bengaluru

(*c*) New Delhi (*d*) Kolkata

19. Which of the following disease is spread by house fly?

(*a*) Typhoid fever (*b*) Cholera

(*c*) Poliomyelitis (*d*) All of these

20. Which oil is used as repellent for mosquito?

(*a*) Neem oil (*b*) Castor oil

(*c*) Citronella oil (*d*) Sesamum oil

21. The intermediate host of malaria parasite is

(*a*) Female *Culex* (*b*) Male *Culex*

(*c*) Male *Anopheles* (*d*) Female *Anopheles*

22. Blow flies belongs to the family

(*a*) Muscidae (*b*) Tepritidae

(*c*) Calliphoridae (*d*) Haematopinidae

23. Which of the following disease caused by infection with *Trypanosoma* species transmitted by the bite of the tsetse fly?

(*a*) Relapsing fever (*b*) African sleeping sickness

(*c*) River blindness (*d*) Bubonic plague

24. Malaria is transmitted by

(*a*) Male *Anopheles* (*b*) Female *Anopheles*

(*c*) Female *Aedes* (*d*) Female *Culex*

25. Which of the following disease caused by the nematode, *Wuchereria bancrofti* and is transmitted by mosquito is?

(*a*) Malaria (*b*) Filaria

(*c*) Dengue (*d*) Plague

26. Relapsing fever is a disease caused by *Borrelia* species of bacteria and transmitted by

(*a*) Mosquitoes (*b*) Red ants

(*c*) Body lice (*d*) Rat flea

27. Which of the following disease is transmitted by mosquitoes?

(*a*) Encephalitis (*b*) Filariasis

(*c*) Yellow fever (*d*) All of these

28. Who first demonstrated the transmission of filarial nematodes by mosquitoes in 1877?

(*a*) Theobald Smith and F. L. Kilbourne (*b*) Sir Ronald Ross

(*c*) Patrick Manson (*d*) Josiah Nott

29. Who first demonstrated the transmission of the agent of Texas cattle fever (bovine babesiosis) by ticks in 1891?

(*a*) Theobald Smith and F. L. Kilbourne (*b*) Sir Ronald Ross

(*c*) Patrick Manson (*d*) Josiah Nott

30. Who first demonstrated the transmission of the agent of epidemic (louse-borne) typhus by human body lice in 1909?

(*a*) Sir Ronald Ross (*b*) Patrick Manson

(*c*) Josiah Nott (*d*) Charles Nicolle

31. Who first demonstrated the transmission of the agent of American trypanosomiasis by kissing (conenose) bugs in 1908?

(*a*) Carlos Chagas (*b*) Sir Ronald Ross

(*c*) Carlos Finlay (*d*) Walter Reed

32. Who definitively demonstrated the transmission of Yellow Fever virus by mosquitoes in 1900?

(*a*) Sir Ronald Ross (*b*) Patrick Manson

(*c*) Carlos Finlay (*d*) Walter Reed

33. Who first demonstrated the transmission of the causative agent of Rocky Mountain Spotted Fever by ticks in 1906?

(*a*) Howard Taylor Ricketts (*b*) F. P. Mackie

(*c*) Carlos Finlay (*d*) Charles Nicolle

34. Which insects are included in the order Blattaria?

(*a*) Cockroaches (*b*) Lice

(*c*) True Flies (*d*) Fleas

35. Which insects are included in the order Phthiraptera?

(*a*) Cockroaches (*b*) Lice

(*c*) True Flies (*d*) Siphonaptera

36. Which insects are included in the order Siphonaptera?

(*a*) Cockroaches (*b*) Lice

(*c*) True Flies (*d*) Fleas

37. What is myiasis?

(*a*) Extreme hypersensitivity to arthropod bites or stings

(*b*) Parasitism by lice

(*c*) Parasitism by fleas

(*d*) Invasion of host tissues by fly larvae

38. What is an arbovirus?

(*a*) A virus transmitted by arthropods

(*b*) A virus found on trees

(*c*) A virus only found in insects

(*d*) A virus only transmitted by ticks

39. How does *biological transmission* differ from *mechanical transmission* by a vector?

(*a*) In biological transmission, the agent undergoes a developmental cycle in the vector before it can be transmitted to another host

(*b*) In biological transmission, the agent does not undergo a developmental cycle in the vector before it can be transmitted to another host

(*c*) In biological transmission, another species of arthropod facilitates transmission

(*d*) In biological transmission, 2 or more pathogens are always transmitted at the same time by the vector

40. What is *entomophobia*?

(*a*) Fear of spiders

(*b*) Fear of mites

(*c*) Skin infestation by arthropods

(*d*) Fear of insects

41. How can a few species of walking sticks (stick insects) cause medical/ veterinary problems?

(*a*) By spraying defensive secretions that can irritate the skin

(*b*) By transmitting pathogenic viruses

(*c*) By transmitting pathogenic bacteria

(*d*) By laying their eggs under the skin

42. Which of these insects often cause inhalational allergies in humans following large emergences of the adult insects?

(*a*) Mayflies

(*b*) Walking sticks (stick insects)

(*c*) Earwigs

(*d*) Praying mantids

43. Which of these blood-feeding insects complete their entire life cycle on the host?

(*a*) Fleas

(*b*) Mosquitoes

(*c*) Sucking lice

(*d*) Kissing (conenose) bugs

44. Which of these groups of arthropods transmit Yellow fever and West Nile viruses?

(*a*) Ticks (*b*) Kissing (conenose) bugs

(*c*) Deer flies (*d*) Mosquitoes

45. Which of these groups of arthropods transmit the causative agents of Rocky Mountain Spotted Fever and Babesiosis?

(*a*) Ticks (*b*) Kissing (conenose) bugs

(*c*) Deer flies (*d*) Mosquitoes

46. Which of these groups of arthropods transmit the causative agents of Plague and murine (endemic) typhus to humans

(*a*) Fleas (*b*) Kissing (conenose) bugs

(*c*) Sucking lice (*d*) Chewing lice

83. How can some millipedes cause medical/veterinary problems?

(*a*) By stinging

(*b*) By spraying defensive secretions that can irritate the skin

(*c*) By transmitting pathogenic viruses

(*d*) By laying their eggs under the skin

47. Sir Ronald Ross is associated with the disease

(*a*) Filarial (*b*) Kala-azar

(*c*) Malaria (*d*) None of these

85. *Vibrio cholera*, the causal agent of cholera, can be carried from one place to another on the feet of flies. This is an example of which type of transmission

(*a*) Obligatory (*b*) Transovarial

(*c*) Mechanical (*d*) Facultative

Fill in the Blanks

1. Causal organism of Dengue fever is transmitted by_________.

2. Causal organism of Chikungunya disease is transmitted by_________.

3. Virus of African horse sickness disease is transmitted by_________.

4. Protozoa causing Babesiosis disease is transmitted by_________.

5. Virus causing Bluetongue disease of cattle is transmitted by_________.

6. Protozoa causing Leishmaniasis disease is transmitted by_________.

7. Causal organism of Dengue fever disease is__________.

8. Causal organism of Chikungunya disease is__________.

9. Bubonic plague caused by bacteria,__________.

10. The causal organism of Onchocerosis disease, *Onchocerca volvulus* is a__________.

Match the Following

1. Match the following disease with their causal group of organism

(a) *Kala-azar*	i) Nematode
(b) Onchocerosis	ii) Protozoa
(c) Chikungunya	iii) Bacteria
(d) Cholera	iv) Virus

2. Match the following insect vectors with their family

(a) Mosquitoes	i) Pulicidae
(b) Bed bugs	ii) Culicidae
(c) Sand flies	iii) Cimicidae
(d) Rat flea	iv) Psychodidae

3. Match the following disease with their vectors

(a) Black fever	i) *Pediculus* spp.
(b) Yellow fever	ii) *Rhodnius* spp.
(c) Trench fever	iii) *Phlebotomus* spp.
(d) Chagas disease	iv) *Aedes aegypti*

4. Match the following disease with their causal group of organism

(a) Malaria	i) Nematode
(b) Filaria	ii) Virus
(c) Yellow fever	iii) Bacteria
(d) Relapsing fever	iv) Protozoa

5. Match the following vectors with their scientific names

(a) Stable fly	i) *Stomoxys calcitrans*
(b) Black fly	ii) *Simulium damnosum*
(c) Eye fly	iii) *Siphunculina funicola*
(d) Sand fly	iv) *Phlebotomus* sp.

6. Match the following vectors with their scientific names

(a) Tsetse fly	i) *Tabanus rubidus*
(b) Horse fly	ii) *Glossina palpalis*
(c) Horn fly	iii) *Hippobosca maculata*
(d) Louse fly	iv) *Haematobia irritans*

Answers

Multiple Choice Questions

1	(b)	14	(b)	27	(d)	40	(d)
2	(c)	15	(b)	28	(c)	41	(a)
3	(c)	16	(d)	29	(a)	42	(a)
4	(b)	17	(b)	30	(d)	43	(c)
5	(a)	18	(d)	31	(a)	44	(d)
6	(a)	19	(d)	32	(d)	45	(a)
7	(d)	20	(c)	33	(a)	46	(a)
8	(b)	21	(d)	34	(a)	47	(b)
9	(d)	22	(c)	35	(b)	48	(c)
10	(b)	23	(b)	36	(d)	49	(c)
11	(a)	24	(b)	37	(d)		
12	(b)	25	(b)	38	(a)		
13	(c)	26	(c)	39	(a)		

Fill in the Blanks

1. Aedes Mosquito
2. Aedes Mosquito
3. Culicoid midge
4. Tick

5. **Culicoid midge**
6. **Sandfly**
7. **Flavivirus**
8. **Virus**
9. ***Yersinia pestis***
10. **Nematode**

Match the Following

1. a) - ii, b) - i, c) - iv, d) - iii
2. a) - ii, b) - iii, c) - iv, d) - i
3. a) - iii, b) - iv, c) - i, d) - ii
4. a) - iv, b) - i, c) - ii, d) - iii
5. a) - i, b) - ii, c) - iii, d) - iv
6. a) - ii, b) - i, c) - iv, d) - iii

www.ingramcontent.com/pod-product-compliance
Ingram Content Group UK Ltd.
Pitfield, Milton Keynes, MK11 3LW, UK
UKHW020232290926
15470UKWH00002B/488

9 789351 309680